P9-DCR-928

# Biology

*by*
*David N. Knowlton, M.A.*

Series Editor
Jerry Bobrow, Ph.D.

INCORPORATED
LINCOLN, NEBRASKA    68501

## FIRST EDITION

ISBN 0-8220-5305-5

574

# CONTENTS

# CONTENTS

# CONTENTS

# CONTENTS

# CONTENTS

# CONTENTS

# CONTENTS

## Introduction

Biology is the study of life. It is a complex and highly organized study that begins with atoms and progresses to the biosphere. Between these extremes are molecules, cells, organs, individuals, populations, and ecosystems. This book will touch quickly on the major concepts of biology. At a glance, the biological world seems infinite in its diversity; however, all living things have basic similarities of structure and function, especially at the atomic level.

## Atoms and Molecules

**Atoms.** The **nucleus** at the center of an **atom** contains protons and neutrons. A **proton** has a charge of +1 and an **atomic mass** of 1. A **neutron** also has a mass of 1 but has no charge. The number of protons in an atom is its **atomic number.** For example, carbon has six protons; therefore, its atomic number is 6, and it is **element** number 6 on the periodic table. The atomic mass of an atom is the sum of the masses of the protons and neutrons in its nucleus. Carbon also has six neutrons in its nucleus, and so the atomic mass of carbon is 12. **Electrons**, which have a single negative charge, are 1/1836th the mass of a proton and are located around the nucleus. An atom is neutral; so, the number of electrons and protons in an atom are always equal. A carbon atom would have six electrons.

The electrons are arranged in **shells** around the nucleus. For elements 1 through 20, shell number 1 can hold up to two electrons, shell number 2 can hold up to eight electrons, and shell number 3 holds up to eight electrons. Because the shells fill from the first shell to the outer, carbon would have two electrons in its first shell and four in its second shell for its total of six electrons. A chlorine atom—atomic number 17—would have two, eight, and seven electrons in its first three shells. A sodium atom—atomic number 11—would have two, eight, and one electrons in its shells.

When an atom's outer shell is full, it is stable. In fact, atoms can gain or lose electrons in order to achieve stability. The sodium atom easily loses its one electron from its third shell. With eleven protons and ten electrons, it has a net charge of +1, making it a charged atom or an **ion**. The chlorine atom becomes a chlorine ion by gaining an electron. Its net charge is then −1.

The number of neutrons in an atom can vary. Chlorine normally has eighteen neutrons to go with its seventeen protons for an atomic mass of 35. A fraction of the chlorine atoms in a sample may have twenty neutrons for an atomic mass of 37. These two types of chlorine are called **isotopes** of each other. All of their other chemical properties are the same; they just have different atomic masses. Carbon with its six protons may have a mass of 12 or, rarely, 14.

Electrons may jump from one shell to another. If an electron **absorbs energy**, it will jump to a higher shell. The amount of energy absorbed equals the energy difference between the shell that it reached and the shell that it started from. In the same way, an electron may **emit energy** by falling to a lower shell. When chlorophyll molecules capture the energy of light, a **photon** of light strikes a molecule of chlorophyll, which absorbs the energy of the photon when an electron in the chlorophyll jumps to a higher energy level. This **excited electron** has energy that can be used to build molecules out of atoms. Instead of falling back to its original shell in one large flash of energy, the excited electron may be passed through a series of molecules that take a little energy at each stop, thereby controlling the release of energy in a cell.

**Molecules.** Atoms may bond together to make **molecules**. One way that they do this is by sharing electrons so that each atom has full shells. An oxygen atom—atomic number 8—has two and six electrons in its first and second shells. Hydrogen—atomic number 1—has one electron in its first shell. If two hydrogen atoms were to each put their electrons next to an electron in oxygen's second shell, then all three atoms would "think" that they had full shells. The *shell model* representation is only two dimensional.

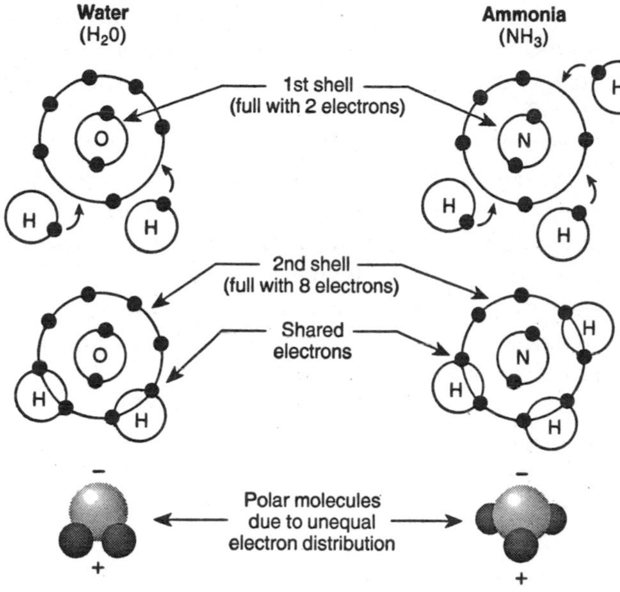


■ Figure 1 ■

Sharing of electrons creates a **covalent bond**. Two similar atoms in a molecule would share electrons equally. An example is a long strand of carbon atoms in which the bond between each carbon is composed of two equally shared electrons. Two different kinds of atoms, as in the case of hydrogen and oxygen, would share electrons unequally making a **polar covalent bond**. This type of bond has a negative and a positive end. In molecules of water and ammonia, for example, the polar bonds give the whole molecule a **polarity**. These molecules are neutral, but they have a negative side and a positive side.

Because opposite charges attract, a sodium ion and a chlorine ion would stick together. In fact, a group of sodium and chlorine ions would stick together. They would not form a molecule because **ionic** bonds do not have a direction; they would form a crystal, also called an **ionic solid**.

A polar molecule, like water, and an ionic substance, like salt, will interact because of their charges. A **solution** can be made by adding salt (the **solute**) to water (the **solvent**). A cluster of water molecules will line up with their positive sides toward each negative chlorine ion and loosen each chlorine from the crystal. Each positive sodium ion will also become surrounded by water molecules. Each water molecule will have its negative side toward the positive sodium ion. The positive sodium and negative chlorine ions drift independently through the solution surrounded by spheres of water molecules. The **polar attractions** described are not bonds but merely attractions, and as such, they are not nearly as strong as a covalent or ionic bond.

Hydrogen atoms make polar covalent bonds with oxygen or nitrogen. Think of two water molecules with one of the hydrogen atoms between the two oxygen atoms. To the hydrogen atom, both molecules look similar, and its electron would form a partial covalent bond with the second oxygen. This phenomenon is called a **hydrogen bond** and is stronger than an attraction but still only a tenth as strong as a full bond.

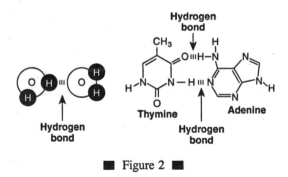

■ Figure 2 ■

Polar attractions and hydrogen bonding between water molecules hold water molecules together. The structure and function of cells depend on water's internal cohesiveness, its temperature-stabilizing ability, and its ability to dissolve many substances. Hydrogen bonding can form a chain of water molecules that reach from the roots to the top of a plant. Hydrogen bonding at the surface of water creates **surface tension.**

Water has an unusually high boiling point because of the polar attractions that hold water together, which gives water a high **specific heat** and allows it to absorb large amounts of heat without big changes in its own temperature. The specific heat of water is 1, which means that one gram of it can absorb one calorie of heat and gain only one degree Celsius of temperature. **Evaporation** causes cooling because molecules absorb energy as they go from a liquid to a gaseous state. It takes 540 calories of energy to make one gram of water molecules evaporate. As water cools, its thermal agitation decreases to the point where hydrogen bonds become permanent, causing water molecules to stand at "an arm's" distance from each other. Water expands when it freezes. Thus, **ice** floats on water forming an insulating sheet over a pond.

**Acids and bases.** A useful definition of an **acid** is that it is a **proton donor.** For instance, an **organic acid** could be a molecule made of a string of carbon atoms with a **carboxyl group** ($-COOH$) on the end. In a water solution, the concentration of hydrogen ions floating around will be low, and so the carboxyl group donates its hydrogen to the solution. The hydrogen's electron stays with the carboxyl group giving it a negative charge, while the hydrogen ion, or proton, has a positive charge. A useful definition of a **base** is that it is a **proton acceptor.** An **organic base** usually has an **amine group** ($-NH_2$) attached. Amine groups accept protons from a water solution that has a high hydrogen ion concentration. Accepting a proton makes the $-NH_2$ into $-NH_3^+$.

An acid/base scale has been derived using water as the base. Water acts as an acid when it gives up one of its hydrogens to become a **hydroxide** ($OH^-$). The hydroxide acts as a base when it accepts a proton to become water. In a sample of pure water, the number of hydrogen ions equals the number of hydroxide, which is defined as a neutral solution. The hydrogen ion concentration of a neutral sample of water is $1 \times 10^{-7}$ moles/liter. A **mole** is a group containing $6 \times 10^{23}$ things. A liter of water would also have the same concentration of hydroxides, $1 \times 10^{-7}$ moles/liter. The acid/base scale is made by taking the negative logarithm of the hydrogen ion concentration, which—in this case—is 7. The symbol for this is **pH** (pronounced "p" "h"). The pH of neutral water is 7. If an acid is added to the water, it may donate protons. If the hydrogen ion concentration rose tenfold to $1 \times 10^{-6}$ moles/liter, the pH would then be 6. If a base were added, it would accept protons from the water. If the proton concentration were reduced tenfold, the hydrogen ion concentration would be $1 \times 10^{-8}$ moles/liter for a pH of 8.

**Organic compounds.** The fundamental atom composing **organic compounds** is **carbon**. Carbon can form large molecules in limitless varieties because carbon has the ability to bond covalently in four directions to other atoms. Carbon may form single, double, or even triple bonds. A **single bond** occurs when two electrons are shared. One line is used in the structural formula. Two lines indicate the sharing of two pairs of electrons, a **double bond**. Three lines represent the sharing of three pairs of electrons, a **triple bond**. A slight modification of the shell model is to visualize the carbon atom sitting in the middle of a **tetrahedron**. The four corners of the tetrahedron are the modified positions of its electrons. Lines drawn from the central carbon to the four corners are its bonding electrons.

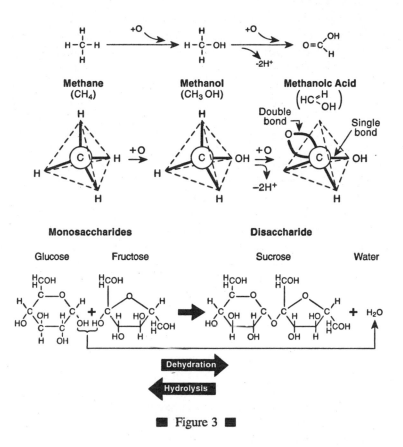

Methane
(CH$_4$)

Methanol
(CH$_3$OH)

Methanoic Acid
$\left(HC\stackrel{H}{<}_{OH}\right)$

Double bond

Single bond

Monosaccharides

Disaccharide

Glucose    Fructose    Sucrose    Water

Dehydration

Hydrolysis

◼ Figure 3 ◼

**Carbohydrates.** **Carbohydrates** are made of carbon, oxygen, and hydrogen and include sugars, starches, cellulose, chitin, and glycoproteins. **Monosaccharides**—like glucose ($C_6H_{12}O_6$) or its **isomer** fructose (also $C_6H_{12}O_6$)—look like pentagonal or hexagonal rings. Two monosaccharide rings may join in a condensation reaction called **dehydration synthesis.** In this joining, a certain **hydroxide** (-OH) group from one ring approaches a certain hydroxide group in another ring. One of the hydroxides and a hydrogen from the other hydroxide break off and join to form water. The oxygen that is left attached to one ring bonds covalently with the exposed carbon of the other ring. The bonding of two rings makes a disaccharide like sucrose ($C_{12}H_{22}O_{11}$). Bonded rings may go through a reverse reaction called **hydrolysis** in which the water is put back in and the rings separate. **Polysaccharides** are formed as more and more rings are added. **Oligosaccharides**, composed of only a few rings, are usually found in association with proteins and are called **glycoproteins**.

Starch is a long **polymer,** or chain of rings in the form of a coil. Individual turns of the coil are attracted to the coils above and below by hydrogen bonds to create a strong structure. Starches are often used by plants to store energy. **Glycogen** is an energy-storage molecule found in animals; it is composed of branched chains. In **cellulose,** the chains lie in sheets, again held together by hydrogen bonding. Cellulose is commonly found in plant cell walls. **Chitin** has atoms of nitrogen in its rings. The hardness of a crab's shell and the stiffness of a fresh mushroom are due to chitin.


**Lipids**. **Lipids,** which include oils, fats, and waxes, are large molecules used for energy storage and to coat a surface to make it waterproof. A large part of the lipid structure consists of long chains of **nonpolar,** covalently bonded carbons. They are **hydrophobic** and do not attract polar water molecules. Organic acids have a **carboxyl group** ( -COOH) at one end. Long organic acids are called **fatty acids.** These acids are commonly chains of sixteen to eighteen carbons. A **glycerol** molecule and three fatty acids combine by dehydration synthesis to make a fat molecule. Much more energy can be stored in a gram of fat than in a gram of starch.

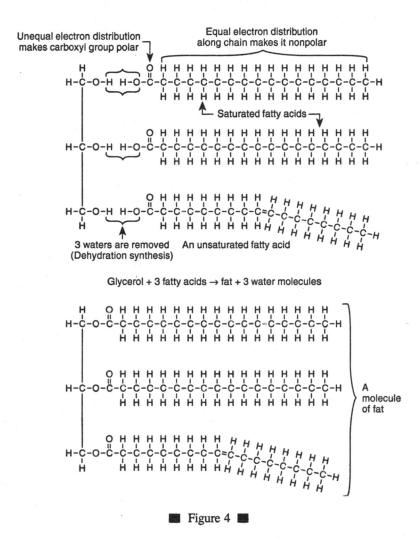

Glycerol + 3 fatty acids → fat + 3 water molecules

■ Figure 4 ■

Stearic acid (eighteen carbons) and palmitic acid (sixteen carbons) make **saturated** fats because all their carbons are single bonded to each other and have the maximum number of hydrogens. Oleic acid (eighteen carbons) has a double bond halfway along its length that gives it a kink and makes it **unsaturated** because it has two fewer hydrogens than it could. Saturated fats can lie together like pencils in a box—that is, with a maximum of surface contact. A weak force of attraction exists between two molecules that is proportional to their surface contact; it is a phenomenon called the **van der Waals force**. The kink in unsaturated fats keeps the molecules from lining up side by side and the van der Waals force from being as effective in holding them together. Saturated fats, therefore, are solids at room temperature because the van der Waals force is stronger—as you see, for example, in bacon grease. The molecules in a sample of unsaturated fats with their weaker van der Waals forces are more likely to be liquids. Liquid lipids are called **oils** and are mostly from plants. Peanut butter is made less oily and more buttery by **hydrogenation**—artificially changing the double bonds into single bonds and adding two hydrogen for each new single bond.

**Phospholipids** are the major component of cell membranes. They are like fat except that one of the three fatty acids is replaced by a phosphate group, creating a molecule that is polar at one end (the negative phosphate group) and nonpolar at the other (the hydrocarbon tails of the fatty acids). Phospholipids are **amphoteric**. The polar end dissolves in water while the nonpolar end does not. **Steroids** are included with the lipids, but they are structurally different from fats. Many hormones start out as steroids. Vitamin D is based on a steroid. **Cholesterol**, found in cell membranes, is a steroid. **Waxes**, the last category of lipids, are long-chain fatty acids bonded to big alcohol molecules or even to carbon rings. They make a waterproof, flexible coating.

**Proteins.** **Proteins**, polymers of **amino acids**, make up hair, muscle, connective tissue, enzymes, and fingernails. An amino acid is a small molecule that has a central carbon covalently bonded to an amine group, a carboxyl (acid) group, a hydrogen atom, and an R-group. About twenty R-groups exist. Dehydration synthesis creates a **peptide bond** that holds two amino acids together, making a **dipeptide.** A longer chain of amino acids is a **polypeptide.** A protein is the total molecule.

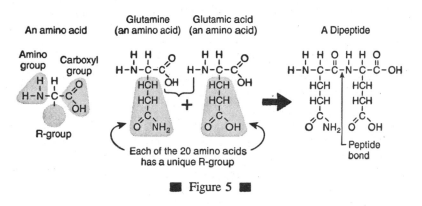

■ Figure 5 ■

The R-groups of amino acids, the key to the diversity of proteins, may be polar, nonpolar, or electrically charged, and they may be capable of hydrogen bonding. Proteins fold into specific shapes during their construction and are held in those three-dimensional shapes by the interaction of the R-groups. A protein strand forms an **alpha helix** that is strengthened by hydrogen bonding between successive loops in the coil. **Fibrous proteins,** which make up the connective tissues of an animal, are composed of coiled or sheeted protein structures. Other proteins, enzymes for example, are **globular.** The sequence of amino acids in a protein is its **primary structure. Secondary** structure refers to whether it is in coiled or sheeted forms. **Tertiary structure** involves its three-dimensional folding. **Quaternary structure** describes the way that subunits of protein fit together to make a larger molecule. Large protein molecules are not stable at higher temperatures. At about 60° C, their structure begins to break down, which is called **denaturation.** A poached egg shows the effects of denaturation.

   **Lipoproteins** are made in the liver and are involved in the transport of fats and cholesterol in the blood. **Low-density lipoprotein (LDL)**

carries cholesterol to sites where it helps form membrane or other material; **high-density lipoprotein (HDL)** picks up cholesterol and carries it back to the liver.

**Glycoproteins**, combinations of protein and oligosaccharides, are found in cell membranes acting in conjunction with the immune system as the basis of self-recognition. They protrude like antlers all over the cell membranes.

**Nucleic acids.** **Nucleic acids** make up the genetic code and control the cell. **Chromosomes** are long strands of nucleic acid (**DNA**) that have been wrapped and folded precisely around proteins called **histones**. A critical feature of chromosomes is their ability to make copies of themselves, an ability that is the basis of growth and reproduction of living organisms. Other forms of nucleic acid, **ribonucleic acid (RNA)**, are able to transcribe the code from sections of the chromosomes, carry this copy out to the cytoplasm of the cell, and construct proteins according to these instructions. This ability allows the nucleus to control the activities of the cell.

The chromosomal DNA, **deoxyribonucleic acid**, consists of four **nucleotides** called adenine, thymine, guanine, and cytosine—abbreviated A, T, G, and C. Nucleotides are composed of a deoxyribose sugar, a phosphate group, and a single- or double-ringed base. The **double helix** structure of DNA was proposed by **James Watson** and **Francis Crick** in 1953 based on x-ray crystallographs made by **Rosalind Franklin**. The sugar and phosphate components hook together by dehydration synthesis to form a strand of nucleotides. The two strands line up looking like a two-dimensional ladder with the rungs formed by the bases reaching each other across the middle. Three rings fit across each rung—for instance, adenine with two rings and thymine with one ring. Now, twist the ladder into a double helix. Each strand of the double helix is a template for the other strand. Adenine on one strand matches a thymine on the other. Guanine is always matched with a cytosine.

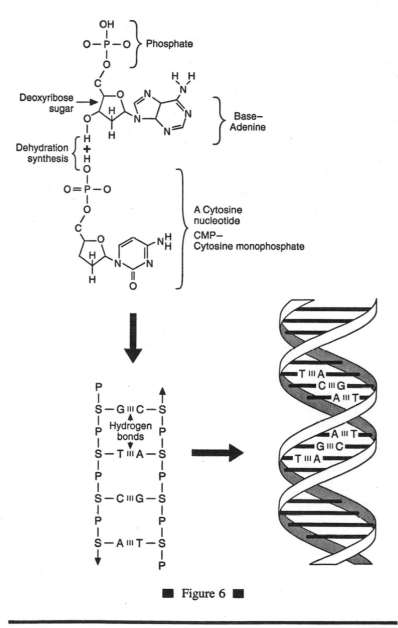

■ Figure 6 ■

Ribonucleic acid (RNA) is found in the nucleus and in the cytoplasm. There are three major differences between RNA and DNA. RNA is single stranded, its sugar is a ribose sugar, and it has a uracil nucleotide replacing each thymine. Studies have shown that RNA conducts the synthesis of protein.

## Biochemical Reactions

**Natural Tendencies.** A chemical reaction has two **natural tendencies**. First, molecules tend to lose energy during a reaction. The energy or heat lost is radiated away. This kind of reaction is called an **exothermic reaction**. A reaction that opposes this trend would absorb heat; it would feel cold. This is an **endothermic reaction**. If molecules are broken up to form smaller molecules with less energy, then heat is given off, and this reaction has followed the first natural tendency. If the reaction forms bigger, more energetic molecules, then the reaction has opposed the first natural tendency. The second natural tendency is for molecules to become more disordered during a reaction. Most reactions involve big molecules with highly ordered atomic structures becoming smaller and with less-ordered molecules.

The two natural tendencies, losing energy and becoming more disordered, are expressed in an equation of **thermodynamics**: $\Delta G = \Delta H - T\Delta S$. The $\Delta H$ refers to the first tendency, which is to lose energy, and $T\Delta S$ refers to the second tendency, which is for the disorder to increase. The $\Delta G$ is the sum of these two tendencies and is called **Gibbs free energy**, an overall measure of whether the reaction will be **spontaneous** and occur on its own or **nonspontaneous** and need help to occur. A spontaneous reaction may have one of the natural tendencies reversed only if the other occurs naturally and strongly enough. For instance, the reaction may absorb heat from the surroundings and get cold only if it becomes extremely disordered. The thermodynamics equation, $\Delta G = \Delta H - T\Delta S$, allows accurate quantification of the tendencies.

**Biological thermodynamics. Biosynthesis**—reaction pathways that result in the construction of large biological compounds—often involves steps that do not follow the natural tendencies and therefore are not spontaneous. Fortunately, means have evolved that allow these unlikely reactions to occur. This process utilizes **coupled reactions**, wherein a spontaneous reaction is coupled with a nonspontaneous reaction so that the net reaction will occur. In cells, the breakdown of **ATP molecules** is strongly spontaneous and couples with reluctant reactions to drive them forward.

**ATP** stands for **adenosine triphosphate**. The adenosine is the same double-ringed adenosine that is part of the DNA molecule. In this instance, a string of three phosphate groups are attached to it. ATP is involved in (1) driving nonspontaneous reactions like protein synthesis and DNA replication; (2) cell contractions, including muscle action and cilia and flagella movements; and (3) passage of materials through cell membranes.

ATP reacts to form **adenosine diphosphate (ADP)** and a **free phosphate group**. In another part of the cell, ADP and P are reconnected to be used again. This recycling of ATP occurs millions of times a second in most cells. The three phosphates joined in a string are not stable. When one breaks off, a more stable molecule is formed; therefore, the reaction is spontaneous.

An example of a coupled reaction is the addition of an amine group to glutamic acid (one of the twenty amino acids) to make glutamine (another amino acid). This reaction is unlikely because it gains energy and has less disorder. When coupled to the hydrolysis of ATP, it becomes possible. An intermediate step occurs when the ATP attaches one of its phosphates to the glutamic acid molecule. The glutamic acid-phosphate complex is at a high energy level and quickly reacts with an amine group to lose some of this energy, which is an exothermic step. The amine joins the glutamic acid and pushes the phosphate off. This process of creating a phosphorylated intermediate is commonly observed in biochemistry. Because the reactant has been destabilized by the addition of a phosphate group, the forward reaction is favored.

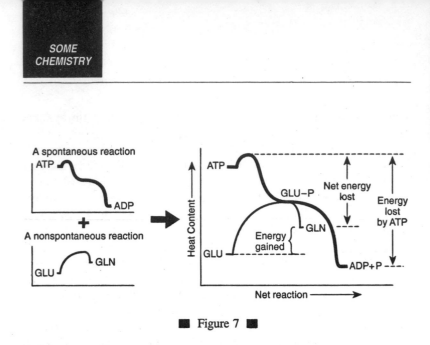

■ Figure 7 ■

**Enzymes.** A flame is a spontaneous reaction once it has started, but it will not start on its own. A barrier exists that must be overcome by an **activation energy.** Adding this energy makes the reactant into an **activated complex,** which contains more energy than the reactant and is unstable. It may change back to its original reactant state (a **reverse reaction**), or it may change to a product molecule (a **forward reaction**) with a lower enthalpy. The cell is able to control the direction of these reactions by producing protein molecules (**enzymes**) capable of lowering the activation energy of a reaction, allowing the reaction to occur at body temperature.

An enzyme is a carefully constructed protein with a three-dimensional shape that allows it to interact with the reactant molecule (the **substrate**). Enzymes are specific. The **active site** of an enzyme is the portion of the enzyme that is complementary to the substrate. In one example of enzyme action, the substrate and enzyme approach one another. The approaching substrate has areas that are polar or ionically charged as well as capable of hydrogen bonding. These areas find complementary areas around the active site that fit only that substrate's arrangement. The areas interact. The active site may be *induced* to change its shape to accommodate the substrate. ATP could add a phosphate to destabilize the substrate further.

Because enzymes are proteins, they are affected by temperature changes. If it is too hot, the protein may denature; the shape of the active site may contort and not accept its substrate. Changes in pH also affect enzyme activity. Functional groups on the amino acids like $-NH_3^+$ and $-COO^-$ are changed. The $-NH_2$ could gain a hydrogen to become $-NH_3^+$ and the $-COOH$ could lose a hydrogen to become $-COO^-$. In both cases, the ionic nature of these groups is changed. If these changes occurred in the active site, the enzyme would be unable to function.

**Inhibitors** may slow or stop an enzyme's activity. **Competitive inhibitors** bind directly to the active site and block the approach of substrate molecules. They may bind strongly, putting an enzyme molecule out of action, or they may bind to the active site weakly, which means that they would be bouncing in and out, allowing the substrate a more or less equal chance. If the decrease in enzyme activity is proportional to the concentration of the inhibitor, then it is a competitive inhibitor. The effects of a competitive inhibitor can be minimized by flooding the cell with substrate so that the inhibitor is crowded away from the enzyme.

An **allosteric enzyme** has more than one active site. If the inhibitor is attached to one site while the substrate is attached to another, then the enzyme is not competing with the substrate. **Noncompetitive inhibitors** are harder to deal with. One kind of noncompetitive inhibitor blocks the active site by sticking next to it; another sticks to the enzyme away from the active site but changes the shape of the enzyme, thus altering the active site. Adding more substrate does not reduce the effectiveness of these noncompetitive inhibitors because they are operating independently of the active site.

## Control

Supplying enzymes when they are needed is a direct method of **control**. Cells can produce enzymes on demand. After their job is done, the enzymes break down, and the cell will not produce any more until they are needed again. But most enzymes do tend to hang around once they have been made. **Allosteric regulators** are molecules that can make an enzyme active or inactive by combining at a site other than the active site. Their binding to the enzyme alters the active site to make it less able to accept a substrate (inhibit) or makes the active site better able to accept the substrate (enhance). **Cofactors** (nonprotein regulators) or **coenzymes** (organic regulators) are active in the process of controlling enzymatic actions. Metallic ions like iron, copper, and magnesium are common cofactors. Vitamins are an example of a coenzyme. These regulator molecules are useful to the cell in controlling rates of reaction along metabolic pathways.

 **Feedback inhibition** is another useful control mechanism. For example, there are cases where the substrate itself is a *positive* regulator. Called **positive feedback**, this reaction occurs when an allosteric enzyme is produced in an inactive state; the presence of substrate will activate it. When the substrate is consumed, the last active substrate-enzyme complex dissociates, and the enzyme returns to its inactive state. On the other hand, the enzyme may always be in its active state. In this situation, the product builds up in concentration. As the concentration of product molecule increases, the product binds allosterically to the enzyme, altering the active site to make the enzyme inactive. **Negative feedback** is a control mechanism in which the active enzyme is turned off by high product concentration. As the product concentration diminishes, the enzyme is activated again as the last product molecule goes away.

 A cell exerts control over its enzymatic activities through structural means. A number of enzymes, each responsible for a single step in a chemical process, are grouped together and held in a membrane. These **multienzyme groups** are found, for example, in the inner membranes of mitochondria where they control cell respiration and the making of

ATP. They are also found in the chloroplasts of plant cells controlling the process of photosynthesis. Essentially, the effectiveness of a multienzyme group is enhanced because the product of one step does not have far to go before it meets the enzyme that controls the next step. Because the product of one step is immediately involved in the next reaction, the forward rate of reaction of the first step is enhanced.

## Introduction

The intracellular chemical processes that allow a cell to grow, reproduce, and do all its other activities are referred to as the cell's **metabolism.** These metabolic processes are divided into **anabolism,** reactions
that consume energy in the production of larger molecules, and **catabo-
lism,** reactions that release or extract energy while breaking down larger
molecules. The two examples that will be mentioned at this point are
**photosynthesis** and **cell respiration. Photosynthesis** is an anabolic
process that takes energy from sunlight to make glucose molecules out
of water and carbon dioxide. **Cell respiration** is a catabolic process
that takes glucose and breaks it down to carbon dioxide and water
molecules.

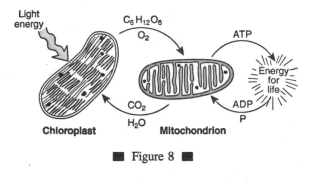

■ Figure 8 ■

## Chemiosmosis X

Reactions in which a spontaneous reaction is coupled with a nonspontaneous reaction have been discussed. This process usually involves
phosphorylating the nonspontaneous reactant so that it becomes unstable

and will then form the desired product. **Phosphorylation** is the transfer of a phosphate from a high-energy ATP to a reactant.

Another way that energy can be transferred is by moving electrons. Reactions involving electron movement are called **oxidation-reduction reactions**. The molecule receiving an electron is *reduced* and the molecule giving up an electron is *oxidized*. The electron involved is at a high energy level and so brings chemical energy with it. For example, a molecule of **nicotinamide adenine dinucleotide (NAD)** can be reduced by two **high-energy electrons** to become $NADH + H^+$. The $NADH + H^+$ can now transfer its electrons to another molecule to energize it. NAD is the primary **electron carrier** in animals.

$NADH + H^+$ carries the high-energy electrons from the breakdown of the sugar to an electron receptor molecule. From the receptor, high-energy electrons pass through a chain of molecules clustered in a multienzyme group that spans a membrane. These molecules are sometimes referred to as an **electron transport chain (ETC)**. In plants, the ETCs are found in the inner membranes of **chloroplasts**; in animals, they are in the inner membranes of **mitochondria**. Each enzyme in a chain is reduced and then oxidized as it accepts and then transfers an electron. The movement of high-energy electrons pulls protons through the membrane. The protons end up being concentrated on one side of the membrane while the electrons, now **low-energy electrons**, are spit out on the other side. The energy lost by the electrons does the work of pumping the protons through the membrane. The ETCs work as **proton pumps**.

A concentration gradient of protons is established. The high concentration on one side now will try to diffuse through to the other side, but the membrane is impermeable to charged particles. The low-energy electrons join an atom of oxygen, which attracts two protons to make water and further reduces the proton concentration on one side of the membrane. The membrane now divides a region of high proton concentration from a region of low proton concentration, creating a great deal of potential energy. The diffusion pressure of the protons causes them to move through specialized molecular channels. These membrane channels use the proton gradient energy to connect phosphate and ADP to make ATP. **Chemiosmosis** is the major source of ATP in living systems.

## Cell Respiration

$$glucose + O_2 \longrightarrow ATP, + H_2O + CO_2$$

**Cell respiration** begins with glucose and oxygen and ends with ATP, water, and carbon dioxide. Cell respiration occurs in and around the mitochondria. Two aspects of cell respiration are that it can be **aerobic**, occurring in the presence of oxygen, or **anaerobic**, occurring without oxygen. Anaerobic respiration is also called **fermentation**. Aerobic respiration is able to extract thirty-six ATPs worth of energy from each molecule of glucose, while anaerobic respiration can extract only two ATPs worth of energy. Cell respiration begins with **glycolysis**. In glycolysis, glucose (a six-carbon sugar) is broken down to pyruvic acid (three carbons) in ten steps. Each step is catalyzed by an enzyme so that its energy can be extracted in small increments and not lost as one blast of heat that might harm the cell.

If oxygen is not present, then pyruvic acid goes through fermentation, becomes a two-carbon compound, and that's the end. No more energy can be extracted. If oxygen is present, however, then the breakdown of pyruvic acid can continue in the **Krebs cycle**, and more energy can be extracted. Here the rest of the high-energy electrons are extracted and carried by $NADH + H^+$ to the inner membrane of the mitochondria.

**Glycolysis.** In *step 1* of glycolysis, a **glucose** molecule is phosphorylated by an ATP. *Step 2* changes the glucose-6-phosphate to fructose-6-phosphate. Another ATP gives up its third phosphate in *step 3* to destabilize the fructose-6-phosphate even more, making fructose-1,6-diphosphate (FDP). The FDP is so unstable that in *step 4*, with help from an enzyme, it splits into dihydroxy-acetone phosphate (DHAP) and glyceraldehyde phosphate (PGAL). In *step 5*, the DHAP isomerizes into more PGAL to continue on. The PGAL is quickly pulled into *step 6* and made into 1,3-diphosphoglyceric acid (DPGA). $NAD^+$ is reduced to $NADH + H^+$ during *step 6* by two electrons given off by PGAL as it is oxidized to DPGA. A free phosphate is also joined to the DPGA at this time. *Steps 6* through *10* all occur twice for each molecule of glucose because each of the two halves of FDP go through. The two DPGAs produced in *step 6* are very unstable. Their breakdown in the presence of the proper enzymes releases a lot of energy—enough, in fact, to charge up two ATPs in *step 7* as the two DPGAs react to become two molecules of 3-phosphoglyceric acid (3-PGA). In *step 8*, an enzyme helps the phosphate shift to the second carbon of 3-PGA to make a pair of 2-phosphoglyceric acids, 2-PGA. *Step 9* sees four hydrogens and two oxygens depart ($2H_2O$), and two **phosphoenolpyruvic acids (PEP)** are formed. As the initials imply, PEP are highly energetic molecules. In *step 10*, each of the PEP becomes pyruvic acid (PYR), charging up two molecules of ATP in the process. That ends glycolysis. The important thing is that the pyruvic acid and ATP have a lot of chemical energy and that the reduced NAD each have two high-energy electrons. Glycolysis can be summed up this way:

$$\text{glucose} + 2ATP + 2NAD^+ + 2P + 2ADP \rightarrow 2 \text{ pyruvic acids}$$

$$+ 4ATP + 2NADH + 2H^+ + 2H_2O$$

**Fermentation.** The pyruvic acid will break down completely only if oxygen is present. Without oxygen, pyruvic acid goes through a couple of steps known as **fermentation.** In plants, fermentation changes the pyruvic acid into ethyl alcohol and carbon dioxide. In animals, fermentation converts pyruvic acid into lactic acid directly. Lactic acid is responsible for the soreness experienced after overexertion. If a muscle does too much, then it may need more oxygen than can be supplied. In this case, the less-efficient fermentation occurs again until the muscle becomes "fatigued." The lactic acid produced under these anaerobic conditions is carried to the liver, where it is further broken down in the presence of oxygen. The panting done after exercise supplies oxygen to the liver.

**Krebs cycle.** In aerobic conditions, pyruvic acid loses a $CO_2$ and becomes an **acetyl group** that is carried through the outer and inner membranes of a mitochondrion by **coenzyme A.** In the **matrix** of the mitochondrion, the two-carbon acetyl group joins a four-carbon compound to begin the **Krebs cycle,** where it is completely broken down. The Krebs cycle is sometimes called the **citric acid cycle.** As in glycolysis, each of the eight steps in the cycle is controlled by a specific enzyme, and each step involves a small change in the energy of the molecule. Because each glucose produced two pyruvic acids, the Krebs cycle occurs twice per glucose.

In *step 1*, oxaloacetic acid combines with the acetyl group brought into the mitochondrion by the coenzyme A, and citric acid is formed. In *step 2*, the citric acid isomerizes into isocitric acid. Three things happen in *step 3*. First, the isocitric acid loses its carboxyl group to make $CO_2$, which diffuses away. Second, the five-carbon compound remaining is oxidized to alpha ketoglutaric acid. Third, this oxidation reduces an $NAD^+$ to $NADH + H^+$. In *step 4*, another $CO_2$ is lost. The four-carbon compound resulting is oxidized by $NAD^+$ making more $NADH + H^+$ and grabbed by a coenzyme A to become succinyl CoA. In *step 5*, the coenzyme A is displaced by a phosphate group that is passed to guanine diphosphate (GDP), making guanine triphosphate that immediately passes the phosphate to an ADP to make ATP. The GDP and the coenzyme A are then recycled, and the succinyl group becomes succinic acid. In *step 6* the succinic acid is oxidized to fumaric acid by transferring two hydrogen atoms to FAD, making $FADH_2$. In *step 7*, a water molecule is added, and the fumaric acid is rearranged to malic acid. And finally, in *step 8*, the malic acid is oxidized to oxaloacetic acid while reducing $NAD^+$ to $NADH + H^+$, and the new oxaloacetic acid is ready to begin the cycle again.

In two turns, the Krebs cycle produces two ATP molecules and sixteen high-energy electrons, which reduce six $NAD^+$ and two FAD to six NADH and two $FADH_2$. The electron carriers, NADH and $FADH_2$, initiate the next phase in the electron transport chains.

**Electron transport chain.** The **electron transport chains** take the sixteen high-energy electrons from the NADH and FADH$_2$ and use them to pump protons into the space between the inner and outer membranes. Eight molecules have been identified in the electron transport chain; they create potential energy in the form of a chemiosmotic gradient. As the protons diffuse back into the matrix of the mitochondrion through the ATP synthases, thirty-two more ATP are

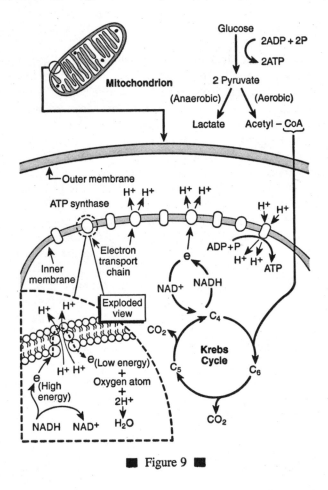

■ Figure 9 ■

produced. The high-energy electrons are now at low energy, and each are accepted by an oxygen atom with two protons to make water. The water eventually reaches a plant where its electrons are bounced to high energy levels again in the process of photosynthesis and the process starts over.

## Photosynthesis X

**Photosynthesis** splits water molecules into hydrogen ions, oxygen atoms, and electrons; oxygen is released. The electrons are energized by light and used in electron transport chains to pump protons. This leads to the production of ATP and the reduction of $CO_2$ to glucose. Recall that respiration is an oxidative process in which molecules of glucose are broken to get their energy. Photosynthesis is a reduction process that converts light energy into electron energy and uses this energy to reduce low-energy carbon dioxide to high-energy glucose.

Photosynthesis occurs in the specialized membranes of the **chloroplasts** found in plant cells, called **thylakoid membranes**. The easiest to spot features are on the **grana**. The grana are like stacks of hollow pancakes; each pancake is called a **thylakoid**. The hollow space in each thylakoid is the **thylakoid space**. The thylakoids are interconnected by tubes through the fluid between the grana. The area between grana is called the **stroma**. The grana membranes contain the chlorophyll and other pigments that trap the energy in certain wavelengths of light.

Photosynthesis is a two-part process. The first part absorbs sunlight, splits water, discards oxygen, produces high-energy electrons, and produces ATP. These steps are referred to as the **light reactions** of photosynthesis and include cyclic and noncyclic photophosphorylation. The other part of photosynthesis, called the **dark reactions**, includes the **Calvin cycle** and uses the ATP and high-energy electrons to reduce the carbon dioxide to glucose. The dark reactions occur in the stroma. In chloroplasts, the principal electron carrier is **nicotinamide adenine dinucleotide phosphate (NADP)**.

**The light reactions.** The **light reactions** are light dependent and begin with the capture of a photon of the correct wavelength. Two types of **photosystems** have been discovered: photosystem I and photosystem II. A photosystem consists of an **antenna** consisting of a few hundred chlorophyll molecules and a **reaction center** with its primary electron acceptor.

**Cyclic photophosphorylation** is the more primitive of the two photosystems; it mostly utilizes the proteins of photosystem I. The electrons are energized by a captured photon, sent through an electron transport chain where they use up their energy to pump protons out of the thylakoid, and then returned at low energy levels back to the antenna. Only ATP is generated by cyclic electron flow using the same chemiosmotic method that the mitochondrion uses. In this case, the energy for the proton gradient comes from light and not from the breakdown of food molecules.

**Noncyclic photophosphorylation** involves both photosystems. Electrons begin by absorbing light energy in photosystem II. The electrons are then accepted at a high energy level and passed through an electron transport chain, which takes some of that energy to pump protons that result in ATP. The electrons now enter photosystem I where they are energized again by another light-capturing event. This time, though, the electrons are shunted to $NADP^+$ reductase, an enzyme that catalyzes the reduction of $NADP^+$ into $NADPH + H^+$. These electrons are now carried off to the dark reactions. Replacement electrons for photosystem II come from the splitting of water, which also releases two protons and one oxygen atom, which joins another oxygen atom to become an oxygen molecule. Noncyclic photophosphorylation produces ATP, NADPH, and oxygen. Most of a plant's photosynthetic efforts involve noncyclic photophosphorylation.

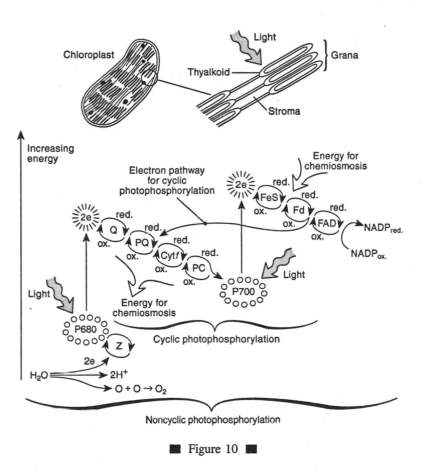

■ Figure 10 ■

**The dark reactions.** The **dark reactions** occur independently of light and consist mostly of a process called the **Calvin cycle,** which occurs in the stroma. Like the Krebs cycle, the Calvin cycle regenerates its starting material in order to go around again. ATP is a source of chemical energy for this cycle, and NADPH brings high-energy electrons to reduce the $CO_2$. The product of the Calvin cycle is glucose. The cycle begins when $CO_2$ is accepted by ribulose biphosphate (RuBP). After three $CO_2$ have been accepted by three RuBP, then six molecules of 3-phosphoglyceric acid (3-PGA) are formed. Absorbing the energy of six ATP, the six molecules of 3-PGA are converted to six molecules of 1,3-diphosphoglyceric acid (DPGA). The six DPGAs are reduced to six molecules of PGAL by six molecules of NADPH. One molecule of PGALs splits out of the cycle to join others in the making of glucose and other sugars. The other five PGALs continue on the cycle, absorbing three ATPs to eventually become RuBP, the starting compound.

An overall equation for photosynthesis is

$$6\ CO_2 + 12\ H_2O - light \rightarrow C_6H_{12}O_6 + 6\ O_2 + 6\ H_2O$$

$$(18\ ATP + 12\ NADPH + 12\ H^+ \rightarrow 18\ ADP + 18\ P + 12\ NADP^+)$$

Observers have noted that the twelve oxygen atoms in the water all end up in the oxygen gas because the twelve $H_2O$ are split in the light reactions, releasing six $O_2$. In the separate, dark reactions, the $CO_2$ is reduced to $C_6H_{12}O_6$ with the production of six $H_2O$. The twenty-four hydrogen atoms from the split water are carried over to the dark reactions by NADPH ($+ H^+$) and become part of the glucose and water formed in the dark reactions. The ATP and NADPH cycle chemical energy and reductive power and are shown in parentheses.

**Photorespiration.** RuBP carboxylase is supposed to accept $CO_2$ to begin the Calvin cycle, but it can also accept $O_2$. In conditions where the $O_2$ cannot diffuse away from the chloroplast fast enough, it may be grabbed by the RuBP carboxylase before $CO_2$ and break the cycle. Under conditions that favor photorespiration—sunny, hot, dry days—as much as 50% of the carbon fixed by the Calvin cycle can be diverted from the production of glucose.

Some plants have "learned" a defense against photorespiration. They are called $C_4$ plants because they incorporate $CO_2$ into a four-carbon compound before injecting it into the Calvin cycle. The majority of plants that do not use this trick are called $C_3$ plants. Sugar cane and corn are two important $C_4$ plants.

A look at the anatomy of a $C_3$ and a $C_4$ plant shows that their structures are different. The veins of a $C_4$ plant are surrounded by **bundle sheath cells**. This ring of bundle sheath cells is surrounded by a ring of mesophyll cells. In the ring of mesophyll cells, phosphoenolpyruvic acid (PEP, which we last saw in glycolysis) accepts a $CO_2$ with the help of an enzyme, PEP carboxylase, to become oxaloacetic acid. This four-carbon compound is the one that gives $C_4$ plants their name (also seen in the Krebs cycle). In two more steps, the $CO_2$ are transferred to RuBP to begin the Calvin cycle. Because $C_4$ plants pump $CO_2$ directly to the RuBP, any oxygen present is crowded away from the RuBP. This prevents photorespiration and makes the plant better adapted to sunny, dry climates.

## History

In the late seventeenth century, **Anton van Leeuwenhoek** observed microscopic organisms. In 1665, **Robert Hooke** reported to the Royal Society of London that he had seen cells, which compose tissue. **Matthias Schleiden** (1838) and **Theodor Schwann** (1839) concluded that all living things are made up of cells. In 1858, **Rudolf Virchow** proposed that "all cells come from pre-existing living cells." The **cell theory**, which is essential to biology, has three parts: (1) all living things are made of cells, (2) metabolism occurs within cells, and (3) all cells come from pre-existing cells, which opposed the then-prevailing theory of **spontaneous generation**—that living things arose from inanimate matter not from living matter. Finally, in 1862, **Louis Pasteur** disproved the theory of spontaneous generation by showing that it could not occur, even if all the proper ingredients were present, unless organisms in the air were allowed to come in contact with those ingredients. The concept of **biogenesis**, the idea that life comes from life, replaced spontaneous generation.

## Microscopes

The most important tool in the study of the cell has been the **microscope**. Robert Hooke used a compound microscope similar in concept to the microscopes used today, which will magnify to around 1,000×. Modern light microscopes employ polarization and the angle and color of illuminating light to increase their effectiveness. The quality of a microscope is not just a measure of its strength of magnification but also its **resolution**, the ability to distinguish closely spaced lines.

Stains have also been used that selectively color only one kind of material in the cell. The most sophisticated technique of this sort is **immunofluorescence microscopy**. In this technique, antibodies also specific for a single type of molecule are tagged with a fluorescent dye and then washed over a cell. The fluorescent antibodies hook onto their molecule, which is then illuminated for viewing.

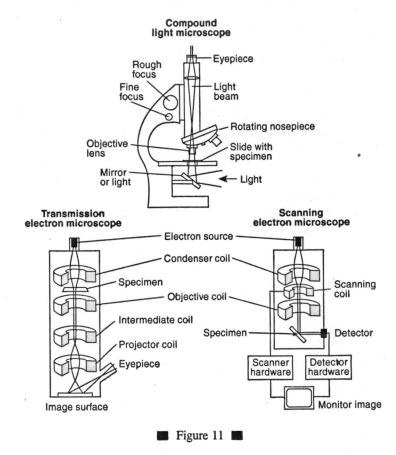

**Compound
light microscope**

Rough focus
Fine focus
Eyepiece
Light beam
Rotating nosepiece
Objective lens
Slide with specimen
Mirror or light
Light

**Transmission
electron microscope**

**Scanning
electron microscope**

Electron source
Condenser coil
Specimen
Objective coil
Intermediate coil
Projector coil
Eyepiece
Image surface

Scanning coil
Specimen
Detector
Scanner hardware
Detector hardware
Monitor image

■ Figure 11 ■

Light microscopes are limited in their resolving power by the wavelengths of light. **Electron microscopes** can resolve 10,000 times better because they use electron beams that have much shorter wavelengths. Instead of glass lenses, an electron microscope uses powerful magnetic fields to bend the electron beam. The surface of a prepared specimen is sprayed at an angle with metal atoms that will reflect electrons.

In a **scanning electron microscope**, electrons are bounced off the coating of metal atoms and into a detector that renders the image on a TV screen. In a **transmission electron microscope**, the specimen is finely sliced and then coated. The electron beam is then sent through this specimen to form an image on a fluorescent screen.

## Origin of Life X

**Formation of the earth.** Scientists speculate that the solar system began to form five billion years ago (BYA) as a swirling cloud of hydrogen and helium with some heavier elements. Computer models of a swirling cloud—given the mass, gravitational effects, size, and torque—do tend to settle into solar systems. As gravity squeezed the forming sun and planets, the compression forces generated great amounts of heat. In the heat and pressure of the sun, fusion began.

On the earth, the heavier elements sank inward and gases formed the atmosphere. As it is today on Venus, the surface temperatures of the ancient earth were probably several hundred degrees. The atmosphere is thought to have been similar to the breath of volcanoes, $H_2O$, $CO$, $CO_2$, $H_2$, $CH_4$, and $NH_3$. In the 1950s, **Stanley Miller** put these gases in a container and excited them with electrical sparks. Checking the residue, he found that amino acids had been created spontaneously. This type of experiment has been repeated many times since; all twenty amino acids have been formed this way as well as the nucleotides, indicating that some tidal pools warmed by volcanoes may have contained a broth of large organic compounds.

Isotopic studies indicate that the earth's crust began to form 4.1 BYA. The oldest rocks are sedimentary and were formed 3.8 BYA. The oldest signs of life have been found in western Australia and southern Africa—fossil bacteria, 3.5 BYA, found in many-layered structures called **stromatolites**. Sheets of bacteria a few inches to a few feet across grow in tidal basins and catch the silt in the water. As they are slowly covered, they grow upward leaving layers of silt and dead bacteria. The fossilized bacterial growths on these stromatolites are very diverse and include photosynthetic bacteria. The chemical systems that are prevalent today had been originated before 3.5 BYA.

**Speculations on the origin of life.** **A.I. Oparin** in the 1920s suggested that life began on the earth spontaneously. At first, it fed on the organic soup that he thought was present and then developed the means to make food on its own. This concept was not well received because the idea of spontaneous generation had been thrown out only decades before by Louis Pasteur. Oparin's ideas did catch on as experimental support, most notably Miller's experiment, developed. Given what is known of life, it is reasonable that four stages were passed through on the path from inorganic matter to living matter.

The first stage is the accumulation of amino acids, nucleotides, and simple carbohydrates. The second stage is the polymerization of the amino acid, nucleotide, and carbohydrate monomers, which has been done in the lab. The amino acid polymers formed are called **protein-oids**. The third stage is the collection of the polymers into clusters or even cell-like structures. When mixed in water, proteinoids form **microspheres**, which show some of the characteristics of a selectively permeable membrane. Oparin had shaken certain mixes of solutions of carbohydrates and proteins and gotten **coacervates**, cell-like structures. If enzymes were incorporated that could break down large organic molecules into microspheres and coacervates, then these **protocells** could grow and compete. The fourth stage is the development of a system of replication.

**Heterotrophs versus autotrophs.** Oparin suggested that the first cells to absorb the organic soup around them were **primitive heterotrophs**. The energy that they needed came from fermenting the large, organic polymers made from the gases in the atmosphere by lightning and the unblocked ultraviolet light.

At some time before the large molecules were used up, cells must have found a way to make their own food. Cells that can do this are called **autotrophs**. The most common autotrophs today use photosynthesis; the rarer autotrophs use chemosynthesis and live off sulfur compounds spewed out of volcanic vents. The photosynthetic autotrophs produce oxygen that is poisonous to primitive heterotrophs. As more and more oxygen was put into the atmosphere, some drastic changes must have occurred. For example, geologists have found vast deposits of iron oxide (rust) which originated about this time. And 0.5 BYA fossils with hard shells containing oxygen appeared. During the time between about 4 BYA, when life is thought to have originated, and 0.5 BYA, when hard-shelled animals left their fossils, the earth's atmosphere changed from a reducing atmosphere that encouraged the formation of life to an oxidizing atmosphere like today's.

**Advanced heterotrophs** must have come along as the oxygen content of the atmosphere grew. These new heterotrophs used oxygen in their metabolic processes to extract a lot of energy from large molecules. The process of **cell respiration** is much more efficient than fermentation. These new heterotrophs had the energy to pursue food actively and to defend themselves.

**Prokaryotes versus eukaryotes.** By the best estimates, life has existed on the earth for 4 billion years. During more than 3 billion of those years, bacteria was the only kind of life. The cell theory suggests that all cells today can be traced back to these most primitive, **prokaryotic** cells (bacteria). Prokaryotic cells do not have a nucleus or organelles. Instead, their DNA are circular and occupy a section of the cell; they do have a cell wall around their membranes, and they often have **flagella**.

Prokaryotes are much smaller than **eukaryotes**, which have a nucleus and organelles. Their nucleus is enveloped by a second membrane that is continuous with the endomembrane system. All multicelled plants and animals and all the **Protista** and **Fungi** are eukaryotes.

The diversity of the prokaryotes that have survived is impressive. Bacteria can live in sulfur hot springs and at the bottom of oceans. They have three separate ways of accomplishing photosynthesis using red, brown, and green pigments. Life would not be possible without the contribution that bacteria make to our ecosystem. Bacteria, the surviving prokaryotic cells, are placed in the phylum **Monera,** which has two sections, **Archaebacteria** (including the ancient type which today survive in hot springs, bogs, and deep sea vents) and **Eubacteria** (including photosynthetic bacteria, spirochetes, and gram positive bacteria).

**The endosymbiont theory.** This theory suggests that prokaryotic cells moved into primitive eukaryotic cells to live symbiotically in them. The main proponent of the theory is **Lynn Margulis,** who has found and organized a great deal of supporting evidence. For instance, chloroplasts and mitochondria have a curious double membrane; the outer membrane is similar to the membranes of the cell that it is in. The inner membrane is different and is similar to the membranes of living prokaryotes. Mitochondria and chloroplasts have their own DNA and manufacture many of their own proteins. When mitochondria and chloroplasts divide, they mimic the binary fusion process used by prokaryotes. There is evidence that the flagella in eukaryotes also came by this route. Bacteria have been found that closely resemble mitochondria, chloroplasts, and flagella.

## Cell Organization X

**Cells** are small (10-30 micrometers) for good reasons. First, the circulation of materials within them is limited to diffusion, good only for very short distances. Second, the membrane is a thin, oily layer interwoven with proteins that wouldn't have the strength to stay together if it were larger. The largest cells are the yolks of eggs that in spite of a great strengthening of the yolk membrane are still fragile. Third, cells depend on necessary materials coming in from the outside and unwanted materials moving to the outside. The amount of this movement is dependent on the surface area of the cell membrane. A cell needs a lot of surface area for exchange with the outside but only a small volume so that things can travel around inside. The surface area to volume ratio must be large for a busy cell. Metabolically active cells are therefore small. Storage cells are relatively inactive and can be much larger. The shape of cells is often a function of their job: a storage cell can be spherical; a nerve cell has long extensions.

The early microscopists called the objects within the cell **organelles,** assuming that they function like the organs of an animal. The **nucleus** was the first to be seen. Surrounding the nucleus is **cytoplasm.** A membrane that has control over what enters and exits the cell encloses the cytoplasm. Organelles allow the cell to get food, process food, eliminate wastes, make new materials, move, and reproduce. All cell activities happen at once, seemingly independent of one another, yet coordinated into a single purpose. The study of these processes is a fascinating pursuit.

## The Cell Membrane (X)

The function of the **cell membrane** is to hold the cell together, to separate the inside from the outside, and to control the passage of materials. The main component of a membrane is the **phospholipid bilayer.** The polar "head" of each phospholipid is aimed away from the center of the membrane, and the nonpolar "tail" reaches to the middle. The addition of unsaturated, bent, phospholipids makes the membrane more fluid. Cholesterol in the membrane makes it less fluid because it helps bind the tails together. Floating in this fluid membrane are **integral proteins.** The portion of the protein that is in the membrane is nonpolar, and the portions that extend outside are polar. Some proteins have a wide nonpolar belt and are polar at each end. These proteins may extend through the membrane protruding at each end. Next to the membrane are **peripheral proteins.** Proteins extending to the outside of the cell are usually **glycoproteins**; they have projections of oligosaccharide chains from their outer ends. There are also **glycolipids** that extend their polar ends away from the cell's exterior, which is called the **glycocalyx**. The projecting chains of the glycocalyx function in cell recognition. Antibodies join to and help destroy cells that they do not recognize. The glycocalyx is also responsible for **contact inhibition,** causing a cell to stop growing when it comes in contact with another cell. This complex visualization of the cell membrane is called the **fluid mosaic model**.

## Passage Through the Membrane $(X)$

**Passive transport.** Ionic particles and larger molecules of all kinds are unable to pass through a membrane. The smaller molecules like oxygen, water, carbon dioxide, and a few others get through by simple **diffusion**, movement of solute molecules from a region of greater concentration to a region of lesser concentration.

The chemical composition of the inside of a cell is different from the outside. If the membrane were not present as a barrier, then diffusion would quickly equalize things. **Facilitated diffusion** often utilizes integral proteins which form **channels** through the membrane shaped to allow only specific materials through the membrane. Integral proteins that aid the passage of materials through the membrane are called **permeases**. Other kinds of facilitated diffusion may be possible. **Carrier molecules** within the membrane may surface on one side, grab a molecule, and carry it through the membrane to the other side. An example of **cooperative transport** is when sodium ions concentrated outside the cell cannot pass into the cell until certain permeases accept them at the outer surface and a glucose molecule also joins. Now, with both a sodium ion and a glucose involved, this permease moves both sodium and glucose through the membrane to the inside. The large diffusion pressure of the sodium is used to pump glucose into the cell. This permease then cycles back to its original condition. Another example of facilitated diffusion involves a **gated channel** that is closed until a trigger molecule opens the gate to allow certain molecules to diffuse through the channel. An "untrigger" molecule closes the gate by removing the trigger.

**Osmosis.** **Osmosis** is the diffusion of materials through a selectively permeable membrane. Imagine that on the outside of a sack made of a selectively permeable membrane there is pure water, while inside the sack there is a solution of sugar in water. Water molecules pass freely through a selectively permeable membrane, while the glucose molecules, which are bigger, do not. Because the water molecules are more concentrated on the outside, they pass into the sack faster than they leave. The sack expands. The expansion continues until the pressure created by the elasticity of the membrane increases the outward flow of water to equal the inward flow, or the sack breaks. The **osmotic pressure**, which is the pressure required to stop the increase of water in the sack, is a measure of the **osmotic potential,** the potential of water to move across a membrane—the greater the **osmotic concentration** (the solute particle concentration), the greater the osmotic potential.

A **hypertonic cell** has more dissolved material inside, and less water, than does the fluid outside. The **osmotic gradient** goes from a lower water concentration inside to a higher water concentration outside; water moves in and the cell expands. In an **isotonic** solution, the cell's solute concentration equals that of the solution so there is no net water gain on either side of the membrane. If the cell is **hypotonic**, it has less solute in it and more water than is outside. It will shrivel as the water diffuses out.

**Active transport.** To get some molecules through the membrane, it is necessary to expend energy and actively pump them through. A good example of this is the **sodium-potassium pump.** The permease responsible moves three sodium ions against their osmotic gradient to the outside of the cell and two potassium ions, also against their concentration, to the inside of the cell. To start this process, three sodium ions from the inside of the cell bind to the permease. This changes the conformation of the permease to allow it to be phosphory-lated by an ATP. The phosphorylation then causes the permease to open to the outside of the cell and dump the sodium. Without sodium and now facing the outside, the permease is changed so that it can accept two potassium ions. Accepting potassium causes the phosphate to leave and the permease to dump the potassium inside the cell. The permease is now ready to go through the cycle again.

**Endocytosis and exocytosis. Endocytosis** occurs when a large object on the outer surface of a cell is brought into the cell and the cell mem-

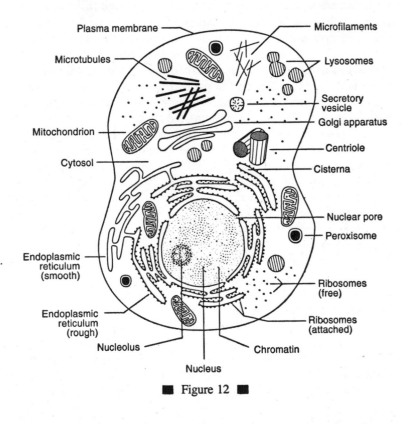

Plasma membrane — Microfilaments
Microtubules — Lysosomes
— Secretory vesicle
— Golgi apparatus
Mitochondrion — Centriole
Cytosol — Cisterna
— Nuclear pore
— Peroxisome
Endoplasmic reticulum (smooth)
— Ribosomes (free)
Endoplasmic reticulum (rough) — Ribosomes (attached)
Nucleolus — Chromatin
Nucleus

■ Figure 12 ■

brane closes over it, creating a bubble—a membrane-bound vesicle—in the cytoplasm. Two forms of endocytosis exist. The first is called **phagocytosis**. In the amoeba, for instance, a pseudopod is pushed out and around the food to engulf it, which forms a **food vacuole** in the cell. **Pinocytosis** occurs when small particles stick to a receptive portion of the membrane, which then sinks inward until the membrane closes over it to form a vesicle within the cell. **Exocytosis** is the opposite process. Material is carried in a vesicle to the cell membrane; the vesicle fuses with the membrane and turns inside out as it becomes membrane; and the contents are spewed out.

## Cell Organelles

Several organelles are composed of membrane; they are part of an **endomembrane system**. These organelles are the rough and smooth endoplasmic reticulum, the Golgi apparatus, the lysosomes, the vacuoles, and the microbodies.

**Rough endoplasmic reticulum.** The **endoplasmic reticulum (ER)** is an elaborate labyrinth of tubes and sacks that accounts for almost half of the membrane of a cell. The **cytosol** is the fluid outside the ER, and the **cisternal** fluid is on the inside. The ER acts as a communication network between the nucleus and the other organelles in the cytoplasm. It is called *rough* when it is dotted on the cytoplasmic side with **ribosomes**, large protein complexes involved in protein synthesis.

The nuclear membrane appears to be a continuation of the rough ER. As the ribosomes make proteins, these proteins are inserted into the ER membrane, which helps them assume their final shape. Some of the proteins imbedded in the ER membrane may act as enzymes for the metabolic processes occurring in the cytoplasm. The great surface area of the ER provides a great deal of contact with the cytoplasm. Other proteins made here enter the cisternal space to become excretory proteins, and some of the integral proteins may find their way to the plasma membrane surrounding the cell. New membrane is made in the rough ER for the rest of the cell and passed on to the smooth ER, then to the Golgi apparatus, and then as membrane-enclosed vesicles to where it is needed.

**Smooth endoplasmic reticulum.** **Smooth ER** is continuous with rough ER; it just lacks the ribosomes. The cisternal proteins and proteins integral with the rough ER membrane diffuse into the smooth ER. The smooth ER's proteins are involved in the production of lipids, carbohydrates, and steroids. For example, in order to produce steroids, adrenal gland cells are thick with smooth ER. Liver cells have a lot of smooth ER to make enzymes to break down drugs and poisons.

**Golgi apparatus.** The **Golgi apparatus** looks like a stack of several hollow pancakes. Vesicles (bubbles) form on the smooth ER and break off. They contain the proteins modified in the rough ER both in the cisternal space and integral in the membrane. These vesicles find their way to the Golgi apparatus and merge with it. The vesicle membrane with its integral protein adds to the Golgi's membrane, and the content of the vesicle is added to the lumen of the Golgi apparatus. Here, the proteins are stored, processed, and made ready for excretion. Each "pancake" seems responsible for doing one stage of the processing before shipping it via vesicles to the next. Vesicles, which may be secretory vesicles that carry new membrane and protein to the plasma or which may be lysosomes, form off the edges of the Golgi apparatus. These secretory vesicles fuse with the inside of the cell membrane, their contents are secreted (dumped), and their integral proteins become part of the cell membrane.

**Lysosomes.** **Lysosomes** are membrane-enclosed bodies that store powerful digestive enzymes made in the rough ER. In some cells, they may fuse with food vacuoles formed by endocytosis. White blood cells engulf bacteria-forming food vacuoles that then fuse with lysosomes.

The bacteria are digested, and the enzymes in the lysosome break down any macromolecules and nonfunctioning organelles that they encounter in the cytoplasm and safely secrete the digested products for reuse. This process helps the cell renew itself. On the death of the cell, lysosomes break the cell down into reusable pieces.

**Microbodies.** Lysosome-like organelles are called **microbodies** and contain specialized sets of enzymes for specific cell reactions. One type of microbody, called a **peroxisome,** contains enzymes that are involved in cell respiration, detoxification of alcohol, and removal of amine groups from amino acids. Peroxisomes often produce hydrogen peroxide, which they quickly change to water and $O_2$. **Glyoxysomes** are another type of microbody that contains enzymes that change fats to sugar in germinating seeds.

**Nucleus.** The **nucleus** is the control center of the cell. It contains the **chromosomes,** the genetic instruction book for the organism, and coordinates the cell's activities. Also visible in the nucleus are the **nucleoli.** The nucleoli are loaded with DNA and proteins and make the ribosomes that move out through the pores to lodge on the rough ER. The RNA (ribonucleic acid) that reads the genetic code is made here also. When the cell is not dividing, the chromosomes are a fine tangle of strands called **chromatin.** When the cell divides, the chromosomes are carefully folded into large, visible structures. Each of the pores in the nuclear membrane are ringed by eight proteins that control the passage of large molecules. The genetic code (DNA) of the chromosomes is transcribed by pieces of RNA that travel out through the pores to the ribosomes where they cause amino acids to join in a specific order to make proteins. The double membrane of the nucleus is continuous with the ER.

**Mitochondria.** The **mitochondria** are the powerhouses of the cell. In them, ATP is produced by the oxidative breakdown of glucose and of cell respirations. The more active a cell is, the more mitochondria it has. Mitochondria have two membranes: the outer membrane seems similar to the cell's membrane; the inner membrane carries the ATP synthase systems and all the cytochrome chain molecules that transport electrons and pump hydrogen. Between the two membranes is the **intermembrane space**, and inside that is the mitochondrial **matrix**. The inner membrane is extremely convoluted, which gives it a large surface area. The folds of the inner membrane are called **cristae**.

**Chloroplasts.** Plant cells have an organelle called a **plastid**, which is not seen in fungi or animals. Colorless plastids, called **leucoplasts**, store starches, oils, or matrix proteins. Colored plastids are called **chromoplasts**. The colors of fruits and flowers are the result of chromoplasts containing **carotenoids**, various yellow-to-orange pigments. **Amyloplasts** are specialized plastids that contain starch and are found in seeds and in the storage cells of carrots and potatoes.

Chromoplasts containing **chlorophyll** are known as **chloroplasts**, a two-membrane structure filled with a protein-rich matrix called the **stroma**. It is in the chloroplasts that photosynthesis takes place. In the stroma are stacks of hollow discs called **grana**, made out of **thylakoid membrane**. The grana contain all the pigments and proteins necessary for trapping light energy. From each grana, the thylakoid membrane stretches in a network of tubules to other grana.

**Cytoskeleton.** The **cytoskeleton** of a cell gives the cell shape and mechanical support. It is composed of **microtubules**, hollow tubes twenty-five nanometers in diameter. These microtubules are made up of two types of protein that spiral around each other to form the tube and act as tracks along which organelles can move. They radiate from a structure near the nucleus called the **microtubule organizing center** and are also found just under the cell membrane in supporting networks.

**Microfilaments** cause cell movement and are made of actin and myosin. The actomyosin complex is a contractile fiber. Extensions from the myosin ratchet their way along the actin filaments to shorten the fibers, using energy provided by ATP.

**Flagella** and **cilia** are composed of microtubules. Flagella move in an undulating fashion to propel a cell. A swimming sperm cell is an example. The smaller cilia are much more numerous on the cell's surface than flagella and act more like oars with a power stroke followed by a recovery stroke. These movements all use energy provided by ATP. Both cilia and flagella are anchored to the cell membrane by **basal bodies**. **Centrioles** look like flagella and cilia in cross section and may actually be migrated basal bodies. Centrioles in pairs take up a station near the nucleus. During the division of the nucleus, centrioles move to opposite poles, and during cell reproduction, they appear to help with the separation of the chromosomes. This role of centrioles, however, is in question. Plant cells do fine without centrioles, and animal cells whose centrioles have been removed still divide.

## Cell Reproduction

The **cell cycle** traces the life cycle of a eukaryotic cell. After a cell is formed, it goes through a growth phase ($G_1$). Then, the cell duplicates its DNA during the S phase followed by another growth phase ($G_2$). The cell cycle described thus far is called **interphase**. Things get interesting during the **mitotic phase**, or during **mitosis,** when the nucleus is duplicated, pulled apart, and the cell divides, a process called **cell division**. This phase is subdivided into prophase, metaphase, anaphase, and telophase and concludes with cytokinesis, the division of the cytoplasm between the two newly formed nuclei.

**Prophase** begins mitosis as the chromatin winds itself up into visible chromosomes. The chromosomes are each composed of two **chromatids** joined by a **centromere**. The chromatids are replications of each other. The nucleoli disappear, and a **spindle** begins to form. The spindle is made of microtubules and associated proteins. The chromosomes move toward the equatorial region between the poles of the spindle, perhaps pushed by the microtubules of the forming spindle.

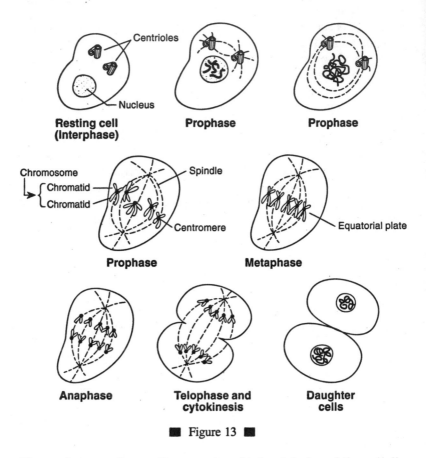

**Resting cell
(Interphase)**   **Prophase**   **Prophase**

Centrioles

Nucleus

Chromosome
Chromatid
Chromatid

Spindle

Centromere

Equatorial plate

**Prophase**   **Metaphase**

**Anaphase**   **Telophase and
cytokinesis**   **Daughter
cells**

■ Figure 13 ■

The nuclear membrane disappears, and microtubules of the spindle
interact with the **kinetochore**, a specialized region of the centromere.
This interaction throws the chromosomes into random motion. **Metaphase** occurs as the chromosomes line up on the equator of the spindle.
Each chromatid of each chromosome is connected to opposite poles by
a spindle ray. (The spindle is symmetrical at this point.) **Anaphase**

sees the separation of the chromatids and the movement of the chroma-tids (now chromosomes) toward their poles. This movement may be the result of depolymerization of the microtubules just behind the kineto-chore, which pushes the kinetochore along toward its pole. At **telophase**, the sister nuclei begin to form. Nuclear membranes are seen around each new nucleus, and the nucleoli reappear as the chromosomes unfold back to the chromatin form. Meanwhile, **cytokinesis** occurs as the cytoskeleton around the equator of the cell begins to constrict, like a purse string, to finally pinch off roughly equal amounts of cytoplasm around each new nucleus, and the daughter cells form.

This process, with two notable exceptions, is similar in plants. Plants do not have centrioles at the poles of the spindle; so, cytokinesis occurs differently. Instead of the daughter cells pinching off from each other, the cells become separated by an **equatorial plate** that grows into a cell membrane between the separating nuclei. This equatorial plate is formed by vacuoles fusing their membranes. When the membranes of each new cell are complete, a cell wall begins to grow between them.

## Discovery of DNA

An unknown material was discovered in 1868 when the protein was separated from the nuclear material. This "new" material contained phosphates, sugar, and nitrogen bases. With a stain developed in 1914, the new material—named **nuclein**—was observable. Microscopists saw that chromosomes doubled and then divided during cell reproduction and that male and female gametes each carried half of the chromosomal material in sexual reproduction. In 1928, work was done with two strains of **pneumococcus bacteria**—a smooth-coated one, which was lethal to mice, and a rough-coated, nonlethal strain. Killing the smooth strain with heat broke it up into little pieces. The pieces of smooth bacteria were not harmful to mice but, when mixed with the rough strain, *transformed* the rough strain into the smooth strain, which *was* lethal to mice. It wasn't until 1944 that the transforming material was discovered to be **DNA.**

Progress was slow because the scientific community was convinced that the genetic code would be found in the proteins of the chromosomes and not in the DNA. By 1952, an experiment by **Alfred Hershey** and **Martha Chase** had finally gotten most scientists to focus on DNA. It was known that proteins contained sulfur but no phosphorous and that DNA contained phosphorous but no sulfur. They tagged growing **bacteriophages**—viruses that "eat" bacteria—with **sulfur** and **phosphorous**, both of which were radioactive. These "tagged" bacteriophages were then turned loose on a fresh bacteria culture where they injected genetic instructions into the bacteria, causing the bacteria to make more bacteriophages. Upon analysis, it was found that the protein coat of the virus containing the radioactive sulfur had stayed on the outside of the bacteria, while the DNA with its radioactive phosphorous had gone into the bacteria. Evidently, the DNA carried the genetic instructions.

Quantitative studies of DNA showed that the amount of adenine and thymine was equal and that the amount of guanine and cytosine was also equal. The nitrogen bases are linked by hydrogen bonding so that adenine and thymine are paired between the strands and guanine and cytosine are paired. (See Figure 6, page 13.) This explosive discovery proved that DNA is the genetic code of life. Watson and Crick (page 12), with key evidence from Franklin, first proposed the double helix structure of DNA and speculated on its methods of replication.

## Replication ✳

During **replication**—the copying of the double helix—the two strands separate so that the formerly matched bases are unmatched. Free triphosphorylated nucleotides, ATP, GTP, CTP, and TTP, approach their complementary bases in each strand. Free adenines pair up with thymines in the unzipped strands, and guanines form H-bonds with cytosines. The two extra phosphates give up their energy as a covalent bond is made between the phosphate of one free nucleotide and a sugar in the newly forming strand. In this way, new complementary strands are made on each of the original strands, resulting in the two double strands.

This replacement process is called **semiconservative replication**. Bacteria grown on a medium containing a heavy isotope of nitrogen $^{15}N$ and then transferred to a medium with the regular nitrogen isotope $^{14}N$ produced a hybrid $^{14}N/^{15}N$ in the next generation. Succeeding generations had all $^{14}N/^{14}N$. The hybrid DNA was the key. The first generation strands of $^{15}N$ had separated, and each built a new complementary strand containing $^{14}N$ on themselves.

**Direction of replication.** The original strand of DNA unzips the hydrogen bonds between its bases to form a **replicating fork**. On each side of this fork, new nucleotides are joined to their complementary

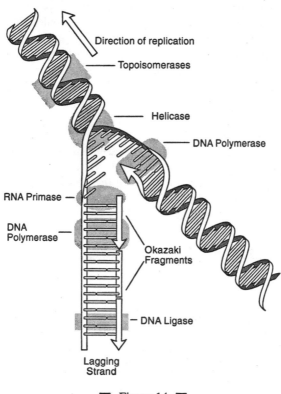

Direction of replication

Topoisomerases

Helicase

DNA Polymerase

RNA Primase

DNA
Polymerase

Okazaki
Fragments

DNA Ligase

Lagging
Strand

■ Figure 14 ■

nucleotides in each strand, but the nucleotides are **antiparallel**. Because the enzymes that do the replication work along the strands in only one direction, one strand—the **leading strand**—will be copied toward the fork and the other strand—the **lagging strand**—will be copied away from the fork in small fragments called **Okazaki strands** (after their discoverer). Because the new strands were being polymerized in one direction, one enzyme—**DNA polymerase**—could handle the attachment of nucleotides to make the new strands. Other enzymes are **helicases,** which do the unzipping of the original strand; **topoisomerases,** which break and reconnect the strands to allow the DNA to swivel as it is replicated; and **ligases,** which reconnect adjacent nucleotides. Apparently, many **origins of replication** exist along a long strand of DNA. At each origin, initiator proteins begin the unzipping, and the replicating forks move away from the starting point to create **replication bubbles.** The replication bubbles run into each other and merge as the replicated strands separate.

**DNA superstructure.** The chromatin seen in the nucleus during interphase is half protein. These proteins **(histones)** help hold the structure of the DNA together. When DNA reaches metaphase, the DNA has condensed by winding two to three times around histone disks to form **nucleosomes.** The nucleosomes pack themselves into a dense spiral about thirty nanometers across. This spiral forms loops, each

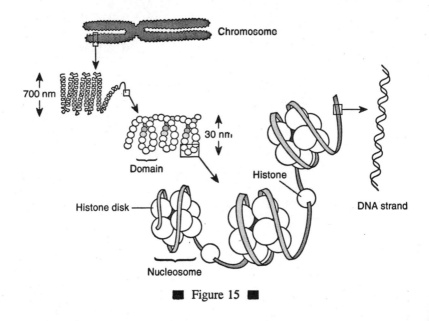

■ Figure 15 ■

loop representing a **domain**. These loops now condense further into a spiral about 700 nanometers in diameter, which forms a section of the metaphase chromosome that is visible under the microscope.

## Protein Synthesis ✳

**The DNA code.** The **central dogma** of molecular biology is the idea that DNA replicates itself and is able to **transcribe** itself (copy segments of DNA with complementary RNA) and **translate** itself (make amino acid sequences—**protein**—from the RNA copies). The DNA code is a sequence of three nucleotides: ATA, CCG, TCA, and so forth. The sequence of triplets in the DNA determines the sequence of amino acids in the protein that is produced.

**Transcription.** **Transcription** is similar to replication except that only one strand of DNA is copied and it is copied in RNA nucleotides. This process begins with **initiation** as the RNA polymerase attaches to a section of DNA called the **promoter**. The DNA of the gene is unzipped; RNA nucleotides find their complementary DNA bases and are joined in a long strand during the **elongation phase** of transcription. The nucleotides of the growing RNA strand peel away as **messenger**

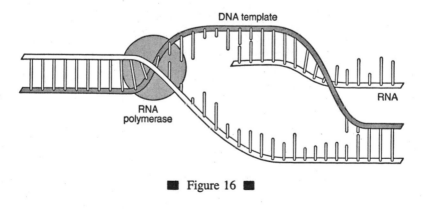

■ Figure 16 ■

RNA (mRNA). During elongation, many RNA polymerases can be working a single gene. Each mRNA strand grows longer as it moves away from the promoter site. At the **termination of transcription,** the growing mRNA strand reaches a **stop codon.** The mRNAs leave, and the DNA strands rejoin each other.

New mRNAs are much longer than they need to be. This observation lead to the discovery in 1977 of **introns,** or *inter*ruptions in the DNA. Between the introns are **exons,** or segments of DNA to be *ex*pressed as protein. Why the gene should be burdened with unneeded intron segments is an interesting puzzle. The more highly evolved an organism, the more introns it carries in its genome. Before any RNA leaves the nucleus, **RNA processing** excises the introns, and the exons are spliced together. Signals for the cutting and splicing are found at the ends of the introns. The large, complex compound of protein and RNA which performs this process is called a **spliceosome.** This process

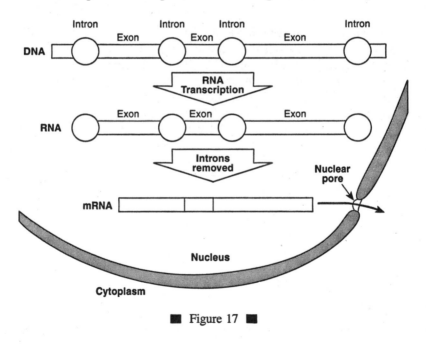

■ Figure 17 ■

may allow the cell to control gene expression. The mRNA, composed of spliced exons, leaves through a nuclear pore to enter the cytoplasm and move to the next phase of protein synthesis.

**Translation.** Genes express themselves by controlling the sequence of amino acids in proteins that are manufactured in the cell. A DNA sequence of AGT will be copied as UCA by the mRNA, bringing in the amino acid serine. The nucleotide triplet (UCA) is called a **codon**. **Translation** is the process by which the cell reads the mRNA and makes amino acid chains—proteins.

**Transfer RNA, tRNA,** are small and roughly "L" shaped. They do their work in the cytoplasm, and their most important feature is that at the bottom of the "L" are three nucleotides, the **anticodon**. The anticodon is complementary to a particular mRNA codon.

A gene may have a thousand nucleotides in its exons to code for a protein. Imagine that at a particular point in the gene the nucleotides AGT may appear. These transcribe to a codon (UCA), which matches an anticodon (AGU) on a tRNA. Now, if this tRNA—with its AGU anticodon—were carrying serine, the connection between the DNA sequence AGT and the amino acid serine would be complete. A key step in translation is for each tRNA to be joined to its particular amino acid. To accomplish this, a group of enzymes called **aminoacyl synthetases,** catalyze the connection of each tRNA with its amino acid. A tRNA carrying its amino acid is called an **activated tRNA**.

In the inactive state, ribosomes are in two pieces; they join together before they synthesize protein. Near the front end of the mRNA is the codon AUG, the **start** codon. **Polypeptide chain initiation** begins when the following all come together: (1) several enzymes—**initiation factors**; (2) guanine triphosphate, which is closely related to ATP; (3) an activated tRNA, carrying the amino acid formylmethionine (fMet); (4) the front of the mRNA; and (5) the small half of a ribosome. The AUG codon and the small part of the ribosome must line up precisely so that the rest of the codons on the mRNA will not be shifted.

 The second part of initiation is to bring in the large half of the ribosome, which has a "P" site (peptide) and an "A" site (aminoacyl).

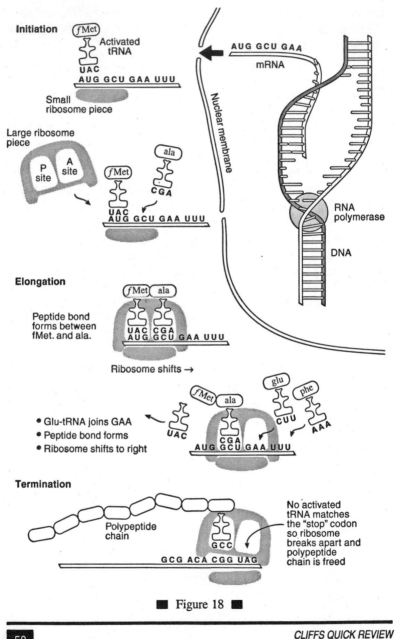

■ Figure 18 ■

Each site can hold an activated tRNA. At the end of the initiation phase of translation, the mRNA is held firmly between the two halves of the ribosome, and the first activated tRNA (anticodon UAC to match the start codon AUG) is held in the P site.

The **chain elongation** phase begins when a second activated tRNA brings its amino acid to the A site and is held there. The amino acid held in the P site joins with the amino acid in the A site. For instance, fMet and its tRNA are held by the AUG codon in the P site. Suppose that the second codon in the mRNA is GCU in the A site. An activated tRNA with the anticodon CGA (carrying alanine) would be held by the codon GCU in the A site of the ribosome complex. The two amino acids, fMet and alanine, would be joined together by a peptide bond. Now, the first tRNA releases its fMet and goes off to be activated again. The tRNA in the A site hangs on to its alanine and shifts to the P site. The fMet is hanging on to the alanine. If the third codon were GAA, then the tRNA whose anticodon is CUU would arrive activated with glutamine. The glutamine would form a peptide bond with the alanine. The alanine tRNA would sail off to be activated with another alanine. The glutamine tRNA would shift over to the P site bringing a new codon into the A site. The growing amino acid chain would be hanging on to the tRNA at the P site. Elongation continues until a *stop* codon (UAA, UAG, or UGA) is reached. At that time, the ribosome splits, and the completed protein is released. This is called **chain termination**.

Ribosomes are often associated with the ER membrane during the elongation phase of protein synthesis. The proteins that they make either are injected into the ER lumen to be released from the cell later as secretory proteins or they become imbedded in the membrane. These integral proteins may find their way through the smooth ER to the Golgi apparatus and by means of vesicles join the cell membrane. **Free ribosomes** do their work in the cytoplasm, and several have been observed moving along a single mRNA all at the same time. This structure is called a **polyribosome**.

## Control of Gene Expression in Prokaryotes ✳

In order to control its own metabolism, the cell uses certain **control mechanisms** for gene expression. For instance, if an enzyme is needed, the gene for that enzyme must be turned on or expressed, and then the gene must be turned off when that enzyme is no longer needed. The control of gene expression is an area of intense study. In bacteria, three control systems have been uncovered: the inducible operon, the repressible operon, and the CAP-cAMP system. Although there are others, these will give you an idea of what's going on.

**Inducible operon.** An **operon** is a group of adjacent genes that are related in their function. Each gene carries the code for a protein and is called a **structural gene**. At the head of an operon is a **promoter** site where RNA polymerase attaches to begin transcription and an **operator** site where a **repressor** protein may attach to disable the promoter. Separate from the operon is a **regulatory gene** that produces repressor molecules.

An example of the **inducible operon** is the lac operon model. Bacteria produce three enzymes to break lactose down into glucose and galactose. Normally, these three enzymes are rare in a bacterium. But when lactose is fed to the bacteria, these enzymes appear. Apparently, the repressor molecule for this operon is normally active; it sits on the operator and blocks any attempts of RNA polymerase to bind at the promoter site. When lactose is added to the medium, the story changes. Lactose binds to the active lac repressor making it inactive. Now, the operator site is left empty, and RNA polymerase can bind to the promoter site and begin transcription of the structural genes in the lac operon. When the lactose is all converted, the repressor is free and active and can block the promoter, shutting the lac operon off. This method of control makes lactose the **inducer**. The addition of lactose induces the lac operon to start working.

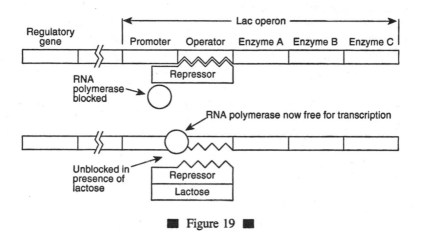

**■ Figure 19 ■**

**Repressible operon.** Bacteria are constantly synthesizing the amino acid tryptophan (Trp). The repressor for the Trp operon is inactive, allowing the RNA polymerase to bind to the promoter site. At certain concentrations of Trp, though, the Trp operon does turn off. This reaction is called a **repressible** control system because the operon is on until something turns it off. As the concentration of Trp increases, it is more likely to bump into the inactive repressor molecule. When it does, it activates the repressor. The now active repressor binds at the operator site, and the Trp operon is shut off. In this example, Trp acts as a **corepressor**. After awhile, the Trp concentration drops, and the repressor resumes its inactive state, allowing the continued production of tryptophan.

**CAP-cAMP system.** The lac operon promoter does not attract RNA polymerase easily on its own; therefore, it makes its enzymes slowly. Thus, this operon does not work rapidly on its own. This problem, the lack of rapid synthesis, leads to a third type of control, which is more

---

positive in nature. The affinity of RNA polymerase for the lac promoter can be increased by a **catabolite activator protein (CAP)**. But CAP is not normally very active. It turns out that **cyclic adenosine monophosphate (cAMP)** can bind to CAP to make a CAP-cAMP complex that sticks to the lac promoter site, boosting the ability of the RNA polymerase to do its job. So, as the lactose concentrations begin to drop and the repressor molecules become more active, the cAMP levels climb. More cAMP means more CAP-cAMP complexes to help keep the lac operon turned on.

## Control of Gene Expression in Eukaryotes

In eukaryotic cells, the control of gene expression isn't as well understood. There are several levels where control is possible: (1) gene rearrangements and amplification, (2) transcription, (3) mRNA processing, (4) transport from the nucleus, (5) translation, and (6) protein modification.

During the cell cycle, chromosomes stretch out to be the chromatin found in the interphase, and then, they are packed and folded into the visible chromosomes seen at the end of prophase. A gene's placement and the manner of this extensive folding may affect a gene's expression. **DNA methylation** causes DNA to be inactive. Methylation of nucleotides occurs during replication. In specialized cells, not all the genes are needed; the unneeded genes tend to be more methylated than the active genes. Studies of *Drosophila* (fruit fly) larva show that **chromosomal puffs** occur in the portions of chromosomes that are more active. The giant chromosomes of fruit flies are special in that they consist of hundreds of copies of the original. The chromosomal puffs are active sections of these chromosomes pushed out where they can actively transcribe DNA into mRNA. As the larva develops, the puffs change their locations.

Because most genes in a eukaryotic cell are not expressed, systems for activating the genes are important. A class of genes called **regulator genes,** similar to the ones discussed in bacteria, has been discovered that produces molecules that may act as activators or repressors. **Promotor regions** associated with eukaryotic genes interact with these molecules. **Gene multiplication** has also been observed. Whole sections of DNA have been found in which a gene is repeated over and over. Interestingly, only 1% of a cell's DNA is typically expressed. Discovering how control mechanisms work in cells is one of the central quests in biology.

## Recombinant DNA Techniques  (X)

History is full of attempts to change the genetic makeup of animals and plants. Corn was a weed several thousand years ago; selective breeding has made it a critical food for the world population. Yeasts used in the production of bread and wine have been carefully selected. Livestock today are bigger and more productive than ever. The list goes on. Recombinant DNA techniques allow us to continue to change genetic makeup by moving genes from one location to another. A plant could be a beet underground and lettuce above ground. The new techniques also allow us to make changes immediately and not to have to wait through generations of coaxing.

A key finding leading to the new field of **genetic engineering** was the discovery of **restriction endonucleases**. This class of enzymes —about 200 have now been identified—act like "molecular scissors" and "cut" double-stranded DNA at specific nucleotide sequences. Many of these restriction endonucleases cleave the DNA in such a way that one strand of the DNA overhangs the other. In this way, endonuclease treatment of DNA can generate many DNA fragments, each with identical ends. DNA fragments with complementary "sticky ends" can be covalently bonded or "pasted" together using an enzyme called **ligase**. These enzymes are not species specific and can be used to fuse DNA from one organism to the DNA from another, the only condition being that the ends of the fragments are compatible. These techniques have been used to modify existing genes, to transfer genes from one

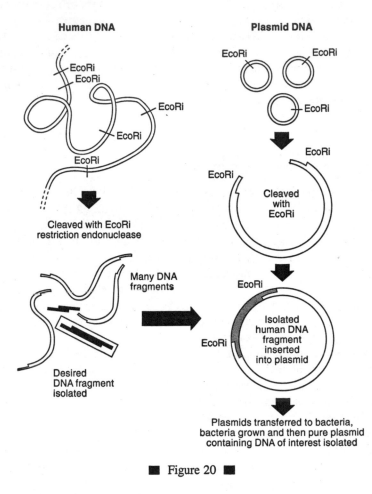

**Human DNA**

EcoRi
EcoRi
EcoRi
EcoRi
EcoRi

Cleaved with EcoRi
restriction endonuclease

Many DNA
fragments

Desired
DNA fragment
isolated

**Plasmid DNA**

EcoRi          EcoRi

EcoRi

EcoRi

EcoRi

Cleaved
with
EcoRi

EcoRi

Isolated
human DNA
fragment
inserted
into plasmid

EcoRi

Plasmids transferred to bacteria,
bacteria grown and then pure plasmid
containing DNA of interest isolated

■ Figure 20 ■

organism to another, and to express genes in tissues in which they are not normally found.

The use of **gene cloning** allows the amplification of identified genes. A DNA library containing all of the expressed genes in a cell can be made by making copies of cytoplasmic mRNA using **reverse**

**transcriptase.** This enzyme, which is produced by a type of viruses called retroviruses (this group includes the human immunodeficiency virus, or HIV, that causes AIDS), can produce DNA from an mRNA template. These DNA fragments, each of which represents the coding sequence of a single protein, can be inserted into bacterial **plasmids,** which are small circular pieces of DNA separate from the normal bacterial DNA. Plasmids can contain genes encoding antibiotic resistance that can be passed easily from one bacterium to another, leading to bacterial strains that are resistant to certain antibiotics. Plasmids containing foreign DNA can be amplified in bacterial hosts, and the plasmid DNA can be easily separated from the normal bacterial DNA, leading to a large, pure supply of the foreign DNA.

This foreign DNA can be used to produce **recombinant protein,** which can be used for therapeutic purposes. For example, the first pharmaceutical product produced by genetic engineering was recombinant human insulin. First, the DNA-encoding insulin was identified and cloned from a library of human DNA. This DNA was then placed into microorganisms that transcribed the DNA into mRNA and then translated the mRNA into protein. The recombinant insulin was then separated from unwanted proteins and used to treat patients suffering from diabetes. Theoretically, any gene can be cloned and used to make large amounts of recombinant proteins using these techniques. This approach has been used to clone and overexpress useful proteins normally produced by the body at low concentrations. Chemicals made by recombinant technology have been used to treat many disorders, including heart disease, cancer, and immune disorders.

The ability to manipulate genes and introduce them into cells has lead to the emergence of **gene therapy** as a possible way to cure patients suffering from illnesses caused by defective genes. Patients can be treated by introducing normal genes into immature bone marrow cells, which can then produce the correct protein. These techniques have also been applied to many other fields, including agriculture where scientists have engineered faster growing trees, tomatoes that ripen but don't spoil, and crops that contain built-in pest-resistance genes, to name just a few examples.

## Mendelian Genetics ✳

**Gregor Mendel** published his work on the genetics of peas in 1866. Although he was a respected clergyman, as a researcher, he was alone—ahead of his time. He kept careful, quantitative notes: He isolated pea flowers in bags to prevent cross pollination; he removed immature stamens from one flower to pollinate another. In his thoroughness, he chose traits that were clear and distinct: seed shape—round or wrinkled; seed color—yellow or green; seed coat color—gray or white; pod shape—inflated or constricted; pod color—green or yellow; flower position—axial or terminal; stem length—tall or dwarf.

**Monohybrid cross.** In a typical entry in his record book, Mendel wrote about testing the cross between a pure yellow-seeded plant, one whose family showed only the yellow seed trait, and a pure green-seeded plant. In his notes, the letter $P$ represented the parent plants. The resulting offspring, the $F_1$ generation ($F$ stands for filial), were all yellow seeded. From this, Mendel concluded that yellow seeds were controlled by a **dominant** factor and that the factor that caused the seeds to be green was **recessive**. Next, he took two of the $F_1$ plants and crossed them. The $F_2$ offspring showed a ratio of three yellow-seeded plants for each green-seeded plant. His notes show 6,022 yellow seeds and 2,001 green seeds. A family tree of the preceding is

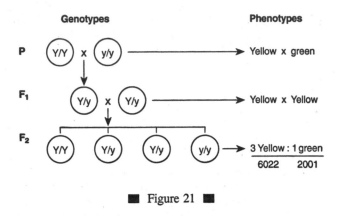

■ Figure 21 ■

Mendel came to the conclusion that each genetic trait was controlled by a distinct factor within each plant. Parents had two of these factors that were separated during reproduction so that a set of chromosomes from each parent was given to each offspring. Mendel's **Law of Segregation** mandated this separation. The factors are called **genes**. When the genes come in two or more forms for a trait, they are called **alleles** for that trait. Mendel's idea was that the pure yellow-seeded parent plants ($P$) had two dominant alleles for yellow ($Y/Y$). They were **homozygous** for the dominant trait. The green-seeded parent plant was homozygous for the recessive green gene ($y/y$). $F_1$ was always **heterozygous** ($Y/y$). The terms *pure* and *hybrid* are also used by breeders. The $F_1$ cross between two parents each hybrid for a given trait is called a **monohybrid cross**.

**Punnett square.** A **Punnett square** is divided vertically and horizontally in half. If parental plants hybrid for seed color are the subject, then the letters $Y$ and $y$ represent their possible genetic contribution. The letters $Y$ and $y$ from one parent are placed above each column, and the letters $Y$ and $y$ from the other parent are placed beside each row. The possible offspring in the four squares take a gene from each parent. Of the four possible kinds of offspring, one is homozygous yellow

(*Y/Y*), two are heterozygous yellow (*Y/y*), and one is homozygous green (*y/y*). In a monohybrid cross, the dominant to recessive **phenotypic ratio** is three to one. **Phenotype** refers to the appearance and **genotype** to whether it is homozygous or heterozygous. Each parent contributes equally; only the genes are dominant or recessive.

**A Monohybrid Cross – Y/y x Y/y**

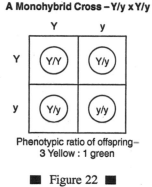

Phenotypic ratio of offspring–
3 Yellow : 1 green

■ Figure 22 ■

To determine if a yellow-seeded plant was pure or hybrid, Mendel used a **test cross**. He crossed the unknown plant, either pure (*Y/Y*) or hybrid (*Y/y*), with a plant possessing green seeds. Green seeds indicated the pure recessive genotype. If the offspring were all the dominant phenotype, yellow, the unknown parent must be homozygous. If the offspring have the phenotypic ratio of one dominant trait for each recessive trait, the parent is heterozygous.

**A Test Cross – Y/?**

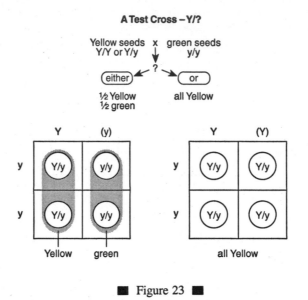

■ Figure 23 ■

**Dihybrid cross.** Consider the inheritance of two traits: seed shape and plant height. Round is dominant, so use $R$ for round and $r$ for wrinkled. Tall is dominant, so $T$ and $t$. Cross a homozygous tall-round parent ($T/T\ R/R$) with a homozygous short-wrinkled parent ($t/t\ r/r$). The gametes (sex cells) of the first parent would carry $T/R$. The gametes of the second parent would carry $t/r$. The $F_1$ offspring can have only one genotype ($T/t\ R/r$), a **dihybrid**, and have the phenotype tall and round. Crossing two of these dihybrid $F_1$s gets interesting. Each $F_1$ can produce the gametes $T/R$, $T/r$, $t/R$, $t/r$. Mendel's **Law of Independent Assortment** covers this. For this **dihybrid cross**, the Punnett square is divided into quarters each way. Across one side put $T/R$, $T/r$, $t/R$, and $t/r$. The same goes on the other side; there are sixteen possible off-spring types. Filling in the sixteen squares shows that the phenotypic ratio is nine $F_2$ with the dominant phenotype ($T/$- $R/$-), three with a dominant-recessive phenotype (*T/- r/r*), three with a recessive-dominant phenotype (*t/t R/-*), and one with the recessive phenotype (*t/t r/r*). The "-" is used to show that either the dominant or the recessive allele could be used, a "?" would indicate that the allele is unknown, and "/" indicates that the genes are on separate chromosomes. The 9:3:3:1 ratio obtained is given only in a dihybrid cross and is therefore a clue to the parents' genotype.

T/T R/R x t/t r/r — The parents, P
T/t R/r x T/t R/r — A dihybrid cross, F₁

|  | T/R | T/r | t/R | t/r |
|---|---|---|---|---|
| **T/R** | T/T R/R | T/T R/r | T/t R/R | T/t R/r |
| **T/r** | T/T R/r | T/T r/r | T/t R/r | T/t r/r |
| **t/R** | T/t R/R | T/t R/r | t/t R/R | t/t R/r |
| **t/r** | T/t R/r | T/t r/r | t/t R/r | t/t r/r |

Phenotypic ratio
{ 9 Tall, Round
3 Tall, wrinkled
3 short, Round
1 short, wrinkled

■ Figure 24 ■

**Partial dominance.** **Partial dominance,** also called **incomplete dominance,** is the case when the heterozygous individual shows a third phenotype, often a blended phenotype. For example, a red snapdragon flower (*R/R*) crossed with a white (*W/W*) will produce pink offspring *R/W*. Partial dominance is easier to work with because there are no hidden traits. The phenotypic ratio resulting from the cross of two dihybrid individuals when considering traits that show partial dominance is 1 red : 2 pink : 1 white.

## After Mendel  X

**Gene interactions.** **Multiple alleles** involve a trait with two or more alleles. In human blood types, for example, there are three alleles: two are dominant, and one is recessive. The result is that a person can have one of four different blood types. The dominant alleles are type A and type B. Geneticists use the symbol "$I^A$" for the dominant type A allele and the symbol "$I^B$" for the dominant type B. The recessive allele is for type O blood ($i$). If an individual has the two dominant genes ($I^A/I^B$), then that person has type AB blood. This is **codominance**. If a person has type O blood, then that person must be homozygous ($i/i$). A type A person could be $I^A/i$ or $I^A/I^A$, and a type B person could be $I^B/i$ or $I^B/I^B$. **Multiple gene inheritance** (also called **polygenic traits**) seems similar to the A-B-O blood types, but there is more blending. Height in people is an example, or skin color, or IQ measurements. All of these traits show a gradual gradation from one extreme to another. This continuous variation is a clue that multigene inheritance is at work. If there are three genes for tallness (there probably are many more), then the tallest person would be A/A B/B C/C, and the shortest person would be a/a b/b c/c. Environmental factors may confuse the results. A person with the genotype A/a B/B c/c would be of an intermediate height.

In the case of **complementary genes,** a single trait is controlled by genes at two locations. As long as a dominant gene is present at each spot, the trait is expressed. But, if the homozygous recessive of either is present, then the trait is not expressed. Zebra finches can be normal, albino with brown-tinged eyes, or classic albino. Crossing the two albino types produces the normal and proves complementarity. Crossing two hybrids produces a 9:7 phenotypic ratio. The 9:3:3:1 ratio that we might expect is replaced by the fact that when a recessive trait shows up, it cancels the other genes. The 3:3:1 ends up being 7.

**Epistasis** is similar to complementarity, but one set does not have equal status. When one set can suppress the other, it is said to be epistatic or "standing upon" (the Greek derivative) the other. The gene for the distribution of coat color in guinea pigs is epistatic to the coat color. It does not matter what color it is as long as it is not distributed. The result of a cross between two dihybrids does not give the 9:3:3:1 ratio but rather a 9:3:4 ratio because only one recessive acts normally. When the epistatic gene is recessive, then it "stands upon" the other and phenotypically looks like the double-recessive condition.

**Collaboration** occurs when two gene pairs interact. A dominant in either produces one trait, a second dominant in both produces a third trait, and a recessive in both produces a fourth trait. Chickens may have four types of combs: rose (*R/- p/p*), pea (*r/r P/-*), walnut (*R/- P/-*), and single (*r/r p/p*).

**Modifier genes** are common because rarely is a trait controlled by only one gene. Even the brown eye/blue eye example used in beginning genetics is affected by modifier genes. Other genes control the distribution of pigments, the tone of pigments, and the amount of pigment. Thus, there are other eye colors besides brown and blue.

**Penetrance** is a measure of the percentage of individuals having a dominant gene who actually show it. Remember that because of epistasis, complementarity, or the effects of modifier genes, having the dominant gene does not mean that an individual will show it. **Expressivity** describes how the dominant trait is expressed—a touch of it or a lot. In webbed fingers, for example, the web may be complete or just appear at the juncture of two fingers.

**Sex determination.** Most organisms are **diploid** and have two set of chromosomes in their nuclei. One set comes from one parent, and the other set comes from the other parent. Mendel recognized this. Having only one set of chromosomes is called **haploid**. Bacteria, some primitive plants, and gametes are haploid. In diploid species where there are separate and distinct males and females, most chromosomes, the **autosomal chromosomes**, have similar partners, but only the pair of **sex chromosomes** are not similar. In humans, the sex chromosomes are called "X" and "Y." Females have a pair of **X chromosomes**, but males have an **X** and a **Y chromosome**. The X chromosome is larger than the Y with more genes on it; thus, human males are haploid for some traits.

**Sex-linked traits**. The genes determining color blindness, a **sex-linked trait**, are on the X chromosome in humans. A female may be pure normal visioned ($X^N/X^N$) or hybrid normal visioned ($X^N/X^n$). A male is either normal visioned ($X^N/Y$) or color blind ($X^n/Y$). Note that the Y chromosome does not carry a gene for this trait. In order for a female to be color blind, she would have to inherit the trait from each parent ($X^n/X^n$). Hemophilia is another human sex-linked trait. Males are either hemophiliacs ($X^h/Y$) or normal ($X^H/Y$). Females can be normal ($X^H/X^H$), heterozygous ($X^H/X^h$), or hemophiliac ($X^h/X^h$). Working out family trees for sex-linked traits is easier with males because males have no hidden traits. **Holandric** genes describe those found only on the Y chromosome. Hair growth on the back of the ear is an example of the effects of a holandric gene.

**Linkage**. Mendel may have chosen genes that were all on separate chromosomes so that independent assortment was possible. In 1900, Mendel's paper was rediscovered, and the similarity between what Mendel described and what the later cytologists observed was noted. Both Mendel and the cytologists agreed that genes sorted independently, gametes were haploid, and fertilized eggs were diploid. In 1906, linkage was established. Sweet peas that had purple flowers and long pollen (*BbLl*) were crossed with those that had red flowers and round pollen (*bbll*). The results should have been purple-long, purple-round, red-long, and red-round in a 1:1:1:1 ratio. Instead, the results were purple-long (*BbLl*), purple-round (*Bbll*), red-long (*bbLl*), and red-round (*bbll*) in a 7:1:1:7 ratio.

In 1910, it was suggested that because the genes were **linked**, on the same chromosome, they could not sort independently. The position of the slashes should be placed between to indicate that *B* and *L* are on the same chromosome: *BL/bl* and *bl/bl*. With this assumption, the gametes from the dihybrid parent should have been *BL* and *bl*. The unexpected gametes—*Bl* and *bL*—occur when the linkages occasionally break to allow a gene from one chromosome to switch with a gene from another chromosome, an occurrence called **cross over**. The site at which cross over occurs is called a **chiasma**.

**Chromosomal mapping.** The more that cross over occurs for a trait, the farther apart the genes for that trait must be. The percentage of cross over that occurs becomes a way of **mapping** the relative positions of genes. A 1% cross over is said to equal a distance of 1 map unit. In the sweet pea example above, the products of cross over—purple-round and red-long—occurred 2/16 of the time, or 12.5%. The genes for color and shape are 12.5 map units apart. Suppose that a gene for height were determined to be 5 map units from the color gene and the height gene were 7.5 map units from the shape gene. A map would have the color gene 5 map units from the height gene, which is 7.5 map units from the shape gene. Detailed maps of some highly studied organisms have been made using this technique to show relative positions of genes.

The **polytene chromosomes** found in the salivary glands of fruit flies are 200 times bigger than normal chromosomes because they have replicated over and over without separating, an example of **gene amplification.** All the component chromosomes lie parallel to each other, and careful staining techniques have revealed bands of different colors  Each band is a gene, and it is possible to see where each gene goes during cell division and during reproduction. The genetic maps based on this cytological evidence put the genes in the same order as the maps from genetic mapping, but the spacing is different. Apparently, cross over frequencies are not the same at any given point in the chromosome. Some places must be more susceptible to cross over than others

**Other chromosomal rearrangements.** In 1951, **Barbara McClintock** published some unusual findings in her work with the genetics of corn. She reported instances of genes actually "jumping" from one site on a gene to another. Recognition for her discovery did not come for thirty years. Like Mendel, she was ahead of her time. These jumping genes, called **transposons,** are important in antibody formation. A related phenomenon is **translocation** in which a segment of a chromosome may move to another nonhomologous chromosome. **Duplication** occurs when a gene moves to the tip of its homologous chromosome, creating a chromosome with two of the same genes. If the broken-off piece of chromosome does not rejoin a chromosome, then it is *deleted*. Without a centromere, a chromosome will not participate in cell division. **Inversion** can occur when a loop forms in the middle of a chromosome and the ends of the loop overlap and switch connections.

Changes in chromosome number are also possible. In gametogenesis, each pair of chromosomes is separated to go into two cells (gametes) containing one of each pair. If the separation of chromosomes during cell division is incomplete, then a gamete may have a pair of homologous chromosomes, a condition called **nondisjunction**. If the gamete is fertilized, then **polyploidy** results. There may be three (**trisomy**) or more chromosomes instead of the normal two, which is not uncommon in plants but is rare in animals. Remember that the other gamete would be missing this chromosome. Most of the time, nondisjunction leads to spontaneous abortions. In humans, trisomy of the 21st chromosome causes Down's syndrome.

**Amniocentesis** is a diagnostic tool useful for determining some unusual gene defects during pregnancy. A hollow needle is carefully inserted into the womb so that amniotic fluid can be extracted. This fluid contains cells from the fetus which may be prepared with certain stains and observed through a microscope for defects. Genetic screening and counseling play an increasingly important role in medicine as our knowledge of and ability to manipulate nature grow.

## Meiosis ✳

Prophase I starts out as in mitosis but soon differences appear. Homologous chromosomes, each composed of two chromatids, move very close together. They line up precisely with their similar genes side by side and with protein cross links to hold them together. The paired, cross-linked chromosomes form a **tetrad** of four strands. Cross overs occur along the chromosomes, and the chromatids become hybrid chromatids.

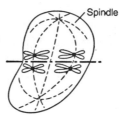

Centriole

Spindle

Nucleus

**Resting cell**

**Metaphase I**

**Anaphase I**

**Telophase I**

**Prophase II**

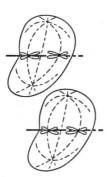

**Metaphase II**

**Anaphase II**

**Haploid Daughter Cells**

■ Figure 25 ■

Approaching Metaphase I, another difference between meiosis and mitosis is seen. The tetrads have two centromeres, and each begins growing microtubules toward the pole that it is facing. These microtubules connect with the microtubules from each pole. The job of this first metaphase is not to separate the chromatids but to separate the homologous chromosomes. This first metaphase is a **reduction** because it reduces the chromosome content from diploid to haploid.

Anaphase I begins as the homologous chromosomes move toward their pole. Telophase I sees the new nucleus forming and an interkinesis (intermission) beginning. Prophase II finds the chromosomes condensing again as they move toward the midline. Metaphase II is more like mitosis. Each kinetochore sends microtubules growing toward each pole so that during Anaphase II the chromatids are pulled apart. In Telophase II, the nuclei form again, but this time the chromosomes are not paired; they are single, haploid.

❋

## Pre-Darwin

**The first taxonomist. Carolus Linnaeus** epitomized the attitudes of the seventeenth through nineteenth centuries. He was convinced that species were fixed and unchanging. Wanting to impose order on the bewildering number of animals and plants, he founded the field of taxonomy by grouping organisms according to physical similarities. It was his idea to refer to an organism by its genus and species names, a practice called **binomial nomenclature** and which is still used today. Thus, a dog is *Canis familiaris*, and a wolf is *Canis lupus*.

**Background to the theory of evolution. Evolution** means change. Used biologically, it means that species are constantly changing to become better adapted. But through the nineteenth century, it was felt that the each species had been made perfectly the first time and that any suggestions that this was not true were antireligious. And in any case, the earth was thought to be only 6,000 years old, hardly long enough for changes to have happened.

　　**Georges Cuvier**, the founder of paleontology, stood against theories of evolution. He found evidence of great upheavals, layers of sedimentary rock twisted upward and downward, which led to his theory of **catastrophism.** Other geologists began to question the evidence in the rocks. **Fossils,** remains or impressions of animals never seen before, were found in **sedimentary rock.** In 1795, it was proposed that geological processes occurring now have always been at work. Remembering that England and Europe were very quiet geologically, it comes as no surprise that this theory, which became known as **gradualism,** stood in direct contrast to catastrophism. Modern geological theory recognizes that the earth is sometimes a very lively place. The modern geologic timetable uses upheavals and great ecological changes as the dividers between the eras and periods.

**Charles Lyell**, a leading geologist of Darwin's time, developed the theory of **uniformitarianism,** which stated that the geological processes occurring now had always been occurring *at the same rate*. With this theory, the age of the sedimentary layers could be estimated at many millions of years by using known rates of erosion and sedimentation.

One of the more comprehensive evolution models was proposed by **Jean Baptiste Lamarck** in 1809. He had arranged a series of fossils from oldest to most recent, showing a gradual change in form, and felt that these changes came from an internal need to achieve perfection. He was impressed with the many examples of how well-adapted animals and plants were to their environments. Lamarck felt that animals exercised the needed parts of their bodies, strengthening some traits and causing others to degenerate. The enhanced traits could be passed on to their offspring. Lamarck's theory held two major points: (1) use and disuse and (2) inheritance of acquired characteristics. For example, he felt that giraffes had developed longer necks because they stretched their necks.

## Charles Darwin

The voyage of the *Beagle* (1831–1836) to chart the little-known coastline of South America created an opportunity for **Charles Darwin**, at age twenty-two, to collect and explore. His notes reveal that shortly after he returned he was already thinking about evolution. Over the next twenty years, Darwin organized a tremendous amount of data for his theory of evolution and thought through its many aspects. When A. R. Wallace came up with the same concept in 1858, friends finally prodded Darwin into publishing. He insisted that Wallace be allowed to share with him a presentation to the Royal Society of London. History credits Darwin, however, because of his prior and extensive work. His book the *Origin of Species* published in 1859 is a landmark in the development of scientific knowledge.

**Darwin's evidence.** **Geology** was a source of evidence for Darwin. He knew how to interpret the sediments and determined that there was ample time for slow and continual evolution. The fossils that he observed suggested to him that animals today had changed from their fossilized ancestors. The oldest layers contained only the simplest animals, but as the layers became more recent, animals appeared to look more like existing animals. Shells were alone in the oldest layers. They were joined later by fish, then by amphibians, then by reptiles, and finally by birds and mammals in the most recent layers. Crudely, geologists were beginning to piece together a table of events for the history of the earth.

Darwin used **comparative anatomy** to show differences and similarities among animals. From comparative anatomy came the knowledge of vestigial organs and embryological development. **Vestigial organs** have degenerated to remnants and are no longer used or needed by the animal possessing them. The presence of a vestigial organ suggests that the animal has changed from having that organ fully developed. Humans have a vestigial tail. Some large snakes have hip and leg bones that are small and unusable. **Embryological** evidence suggests that during their embryological development (their **ontogeny**) humans go through the stages seen in the embryos of simpler species. Early in ontogeny, we pass through stages that remind scientists of the embryos of preceding animals, even at one point of developing gill pouches.

**Homologous structures** were cited by Darwin as suggestive of evolution. The bone structures of the digits, wrist, radius, ulna, and humerus in humans, cats, porpoises, and bats are based on the same structure. Darwin suggested that these animals **diverged** from a common ancestor. Species that do not have a common ancestor may also look alike because they have adapted to similar environments, a phenomenon called **convergence**. Examples are porpoise, shark, ichthyosaur, and fish. All four have adapted to life in the water and look alike, but all began their evolution at separate points. Their fins, tails, and body shape are **analogous structures**. Species with analogous structures are said to have **converged**. A major difference between homologous and analogous structures is in the degree of complexity. Analogous structures show obvious differences on close inspection, while homologous structures agree in detail.

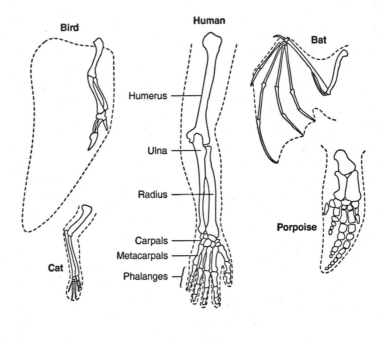

■ Figure 26 ■

**Selective breeding** is another area from which Darwin obtained evidence. Animals and plants could be made to change in appearance over only a few generations simply by careful selection of parents. Terriers, Great Danes, cocker spaniels, and bulldogs have been bred by people interested in bringing out certain traits. Cauliflower, Brussels sprouts, and cabbage have been bred from a wild plant (*Brassica oleracea*) that has little resemblance to any of them.

Since Darwin's time, **molecular biology** has added more evidence. DNA is the universal code of all life on earth. From bacteria to people, DNA does its job in the same way, using the same enzymes. Studies have shown that the more closely related two organisms are, the more their DNA sequences match. **DNA-DNA hybridization** is a process in which DNA from two sources is heated and allowed to unzip, and then mixed and cooled to form hybrid DNA. The hybrid DNA is isolated and then heated until it begins to unzip. The higher it must be heated before it comes apart, the better the two DNA strands match.

Proteins are windows to the DNA. It is possible to analyze amino acid sequences in proteins. Comparing human beta hemoglobin, which has 146 amino acids, with that of a gorilla, there is a difference of one amino acid. Comparing it to that of mice, there is a difference of twenty-seven amino acids. Frogs have sixty-seven amino acids that are different. Graphing these changes against time reveals that their rates of change seem fairly consistent. These molecules are a **molecular clock**. Given the difference in a molecule taken from two related animals, it is possible to estimate when the two diverged along separate paths.


**Darwin's mechanism.** Darwin collected an impressive amount of evidence to support the idea that organisms change over time. What made his theory compelling, however, was his proposal as to how evolution occurred. The driving force behind evolution, Darwin explained, was summarized by five facts:

1. Populations produce more offspring than their environment can support.
2. Physical and behavioral variations of all types exist within these offspring.
3. Offspring compete for the available resources.
4. Those individuals possessing favorable or advantageous traits succeed more often than those with unfavorable traits.
5. Successful organisms have more offspring and can pass on those traits that made them successful.

Over many generations, in a process called **natural selection,** favorable changes would accumulate until change was evident. Darwin did not say how traits originated or how they were passed on. Mendel's work was not available, and nothing was known of how chromosomes worked. Darwin said that nature, in the form of the total environment, "selected" those characteristics that made an organism better able to survive. Another phrase that sprang up was "survival of the fittest."

**The English peppered moth.** Before the industrial revolution in England, the air was cleaner, and trees, rocks, and lichen downwind of cities were not coated with soot. Collections of the peppered moth at that time showed mostly light-colored moths and only a few dark-colored ones. The light-colored moths resting on clean lichen and birch trees blend in well. Their predators, birds, cannot see them easily. However, dark-colored moths are not as well camouflaged and a higher percentage of the dark moths end up as dinner. But the dark moths do have one advantage—when shifting position from place to place during the day, they are harder to spot, while a moving light moth is easy to spot. As pollution darkened the landscape, the light forms of the moth became easier for the birds to see on the soot-covered trees; therefore, they were eaten more often. The dark forms had an advantage, and fewer dark forms were eaten. Then, more dark moths were having more offspring who had a better survival rate. Over the span of many generations, a formerly uncommon trait had became the predominant form.

## Neo-Darwinism

**Adding Mendel's genetics.** Mendel's findings showed how two parents could have a variety of offspring. New microscopic techniques made it possible to see chromosomes during mitosis and meiosis. Crossing over during meiosis was a definite source of variation, and the idea of multigenic inheritance explained continuous gradations in phenotype. Neo-Darwinism has developed as new findings have been fit into Darwin's original concept. A rapidly expanding inventory of fossils and new understandings of ancient ecosystems, discoveries in the molecular processes that control life, and more precise ways of determining relationships among species have all contributed to an improved theory of evolution.

**Population genetics.** The study of **population genetics** allowed scientists to focus on the inheritance of visible traits among the members of a population. Instead of working with individuals as Mendel did, scientists considered the whole population. All of the genes in a reproductive group at a given time are called the **gene pool.** For example, some people can roll up their tongue like a pea shooter, a trait that is dominant (*R/R* or *R/r*). People who can't roll their tongues have two recessive genes (*r/r*). Suppose that a study reveals that 36% of a population are nonrollers. A Punnett square is constructed with *R/-* and *r/-* across the top and *R/-* and *r/-* on the side. This trait is held equally by males and females. Knowing that the offspring in the bottom right corner represent 0.36 of the population and that the contributions from the top and side of the square are equal allows us to conclude that 0.6 of the genes in all the males and females are *r*. Put 0.6 in front of each *r*. This fraction—0.6—is the **gene frequency** of the recessive trait. Because there are two alleles, the gene frequency of both must add up

**Punnett Square of Population Genetics**

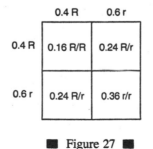

|  | 0.4 R | 0.6 r |
|---|---|---|
| **0.4 R** | 0.16 R/R | 0.24 R/r |
| **0.6 r** | 0.24 R/r | 0.36 r/r |

■ Figure 27 ■

to one, and the gene frequency of the dominant trait is 0.4. So, put 0.4 in front of the $R$s on the top and the side. The rest of the square can now be filled in—0.16 $R/R$, 0.48 $R/r$, and 0.36 $r/r$. One piece of information about the population allowed the whole Punnett square to be filled in.

**The Hardy-Weinberg Law.** The basic concept of this law was developed independently in 1908 by two people and is called the **Hardy-Weinberg Law** after them. They said that in a population that is in equilibrium the gene frequencies would not change over the generations. No change means no evolution. The Hardy-Weinberg Law set up a baseline from which evolution could be measured and gives five conditions that must be met for no evolution to occur:

1. The population must be large, more than 10,000 individuals.
2. There must be no net changes in the genes of the group due to mutations.
3. The population must be isolated, no individuals migrating in or out.
4. The selection of mates has to be random so that each female has equal access to each male and vice versa.
5. All genotypes have an equal chance at success.

Because most populations are large, the first condition is usually met. A large population is necessary if tiny, random fluctuations are going to cancel out. And with a large population, the accidental loss of one or a few individuals will not affect the gene frequency. Suppose that a dozen individuals on the fringe of the main population were somehow pushed into a new area and cut off from the main population. Adapting to fringe conditions, this small splinter group would most likely have a gene frequency different from the main group. Furthermore, if one of these twelve individuals were not as fit as the others, the death of that individual would take out a large percentage of unfit genes. This situation is called **genetic drift**, and it would as likely lead to bad results as to good results.

The second of the Hardy-Weinberg conditions is that mutations are random and rare. The importance of an occasional mutation is that it adds new genes to the gene pool. Because mutations are random and do not shift back to the original as often as they shift away from the original, a **mutation pressure** exists. This pressure would cause a slow drift in gene frequencies, and this drift would be random and not in itself cause directed changes.

**Migrations,** the third condition, have not been a major factor in the many populations that have been studied. As a minor factor, small interbreeding groups do add to the gene pool or are taken from the gene pool. This is called **gene flow** and would upset the Hardy-Weinberg equilibrium. The observations of naturalists suggest that the disturbance is small.

The fourth condition, **random mating**, is almost never seen. In fact, animals go to great lengths to attract and select mates. The success of their reproduction behavior has to do with the fitness and appearance of an individual. And the last of the Hardy-Weinberg conditions, **equal reproductive success**, is also not seen in real populations. A group of offspring will have many small to large differences. It is these variations that make some successful and some not as successful.

**Some causes of evolution.** The fact that not all of the Hardy-Weinberg baseline conditions are met suggests that evolution is inevitable. One of the most important of the factors that steer evolution is **selection pressure**. As an example, reconsider the tongue-rolling ability. Suppose that 1% (0.01) of the population is not able to roll their tongues. Using this data and the Hardy-Weinberg ideas reveals that the gene frequency of $r$ is 0.1 (this is the square root of 0.01), and so the gene frequency of the dominant trait ($R$) is 0.9. Now, further suppose that a selection pressure of some sort is making the tongue rollers less fit so that in each generation the frequency of $R$ drops by 0.1. This population is not in equilibrium; it is evolving. In nine generations, the population would change from having 99% tongue rollers to having only 19%.

Real examples of large changes in populations can be found in nature. The English peppered moths shifted from light to dark forms in just a few decades because of natural selection. In fighting infections, penicillin has been a lifesaver. It has a down side, though. *Staphylococcus aureus*, a pernicious bacteria, has a **preadapted trait**. This trait, caused by a gene or a group of genes, is rare, and while the trait was useless to the bacteria before penicillin, this gene made the bacteria immune to penicillin. It is important to note that the bacteria did not develop immunity *in response* to penicillin; the bacteria was already immune. If it had not been immune, penicillin would have wiped it out. Penicillin created a selection pressure that caused the frequency of this trait to shift from uncommon to common in a short time.

The **bottleneck effect** occurs when a catastrophe wipes out most of a population. The few survivors do not possess all the genes of the formerly large gene pool, and gene frequencies can fluctuate greatly due to genetic drift in the small population. Although, at the turn of the century, northern elephant seals were hunted to near extinction, the slaughter stopped when they became so hard to find that hunting them was no longer profitable. After a time, elephant seal byproducts were less valuable, and elephant seals were declared an endangered species and thus protected by law. Estimates are that only twenty elephant seals lived through to the end of the nineteenth century, but now they number in the 30,000s. A check of their genetic variability was done by examining the variety of alleles at twenty-four positions in their chromosomes. No variations were found. A contrast can be made with the southern elephant seal, which was not hunted to near extinction. In them, great numbers of variations were found. Zoos around the world are involved in saving endangered species and go to great lengths to breed their animals to preserve as much variety as possible.

**Adaptive radiation.** Darwin's finches are a classic example of **adaptive radiation.** A few hardy finches were blown from the mainland of South America several hundred miles across the Pacific to the Galápagos Islands. On the islands, they were isolated from each other in small groups. When Darwin observed them, he found a great variety of finches. They had changed from the original ancestors to become birds that had adopted many new forms, from huge beaks for cracking nuts to wide mouths for catching insects. Many available food sources were utilized by these finches. This scenario is an example of the **founder effect**, a type of genetic drift. The birds who were the founders of new generations of birds on the islands radiated out (adapted) to fill in the available niches in their environment.

**Speciation.** **Speciation** occurs when a group of animals of the same species becomes two species with overlapping or separate ranges. Two animals are members of the same species if they can mate to produce fertile offspring. Animals that are part of the same gene pool belong to the same species. A gene pool may include a great variety of features and be spread over a large area. Within this area, there may be **microclimates**—such as streams, grassy fields, forests, or mountainsides—that experience different temperatures, varying amounts of water, different soil, and varying amounts of sunlight, to name a few. The animals living in these microclimates may be specialized for these local areas, even though they are part of the larger gene pool.

Another aspect of a large range is that conditions may change gradually from one end of the range to the other. Traveling from a field up the side of a mountain, you may experience a gradual change of conditions. Taking plant samples from higher and higher locations up the slope and growing them in the lab yields progressively shorter plants. Continuous phenotypic changes due to gradual changes in geography are called **clines**. The plants at one end of a cline have adapted differently and look different from plants at the other end. In some cases, this difference may be so great that individuals from each end cannot mate. However, even though the extremes cannot mate, they are part of the same gene pool and are still the same species. The genes from one end of the population range may eventually be passed to the other end.

If the cline were cut in half or if the microclimates became isolated, then the animals would no longer be part of the same gene pool.

Variations, even mutations, occurring at one end would no longer be able to spread across the **barrier**. As each group adapted to its own area, it would become distinct from the group from which it was separated.

Squirrels of the Grand Canyon are a good example. Before the Grand Canyon developed, a million years ago, squirrels ranged over the whole area of what is now Arizona. The Grand Canyon divided the range, and today, two distinct kinds of squirrel are found. The Kaibab squirrel, which is redder, lives on the north side, and the silver-gray Abert squirrel lives on the south side—two distinct types on their way to becoming two species. Each group has its own selection pressures. Changes in gene frequencies of the group on one side are not passed across the canyon to the group on the other side.

## Punctuated Equilibrium

Evolution was perceived as being a process that responded to slow and gradual geological changes. Species would, it was thought, very slowly acquire new traits until, at some point, a new animal would develop. The fossil record did not quite agree with this. Fossils of a single species collected in one strata of rock were similar throughout a layer that may have taken millions of years to form, and the next layer would have a different set of animals and plants that were consistent in form throughout that layer. In fact, geologists use the species in a layer to identify that layers from others. That this did not agree with gradualism was overlooked because the fossil record was incomplete, and it was assumed that the missing links would be discovered one day.

Recently, **Niles Eldredge** and **Stephen Jay Gould** have taken a more literal look at the fossil record. What they saw were long periods of stability punctuated by brief periods of upheaval. The lengths of time are in geological terms. The stable periods were tens of millions of years long, and the shorter times were tens of thousands of years long. Their view is that plants and animals underwent intense selective pressures and evolved during the shorter periods while maintaining a state of equilibrium during the stable times. This is called the **punctuated equilibrium** theory. Discussions of which theory is more acceptable hinge on finding more data and on a better understanding of factors that control the rates of change.

## Phylogeny

**Taxonomists**, people who group and name animals, develop family trees of organisms. These trees have tried to depict the evolutionary history, or **phylogeny**, of animals and plants. Several methods have been used to determine which species is most like another or to judge how different two organisms are.

The first, and still the most important, method is a close study of the anatomical characteristics. This method shows how closely or distantly two animals are related. Determining which characteristics are the best to use has always been a healthy source of argument. A knowledge of fossil predecessors and an understanding of embryological development is also important to making distinctions. The decision to place organisms together must also take into account homologous and analogous structures.

A good portion of taxonomy depends on the insights and intuition of the investigator, but two methods have been used that try to make the process more objective, one of which is **phenetics**. In this system, all traits are considered and ranked under the assumption that all characteristics are equally useful in determining phylogeny and that convergent evolution is not a big factor. **Cladistics** is another attempt to be objective. As in phenetics, each trait is weighed equally, and the problem of analogous structures is ignored. Traits are chosen for comparison that are held in common by the animals being investigated. The phylogenic distance of each is then determined. Critics point out that choosing the traits held in common is itself a bit subjective. Although these last two methods have important applications, subjective judgments have always been important.

Molecular taxonomy is a recently introduced method that has great promise. As has been discussed, mutations are relatively rare but seem to occur at a constant rate much like the radioactive decay of some isotopes. DNA acts in a sense like a molecular clock. When DNA samples from two species are compared using DNA-DNA hybridization, the degree of difference is proportional to the phylogenic separation of the two. The time that has passed since they had a common ancestor can be approximated.

## Classification

Almost one and a third million species of animals and plants are known. The taxonomic system started by Linnaeus has expanded to provide a place for each of these organisms. The most similar of organisms are listed in the species level. Several similar species are grouped at the genus level. Above genus is family and then order, followed by class, phylum, and kingdom. The level of kingdom is the most general. A dog is classified as follows, starting with its kingdom and working toward the more specific: kingdom—Animalia, phylum—Chordata, class—Mammalia, order—Carnivora, family—Canidae, genus—*Canis*, and finally species—*familiaris*. All the animals in the phylum Chordata have backbones. The classes in this phylum include sharks, fish, reptiles, amphibians, birds, and mammals. The class Mammalia, which includes dogs, also includes bats, deer, kangaroos, and people. Dogs share the order Carnivora with a group of meat eaters including cats and bears. The family Canidae is known for its canine teeth. Following the convention of binomial nomenclature, dogs are called *Canis familiaris*, their genus and species. Wolves, by the way, are *Canis lupus*. A way to remember the divisions has been the mnemonic device *K*ings *p*lay *c*hess *o*n *f*olding *g*old *s*tools.

The five-kingdom system is the most popular scheme at present. It recognizes Monera, Protista, Fungi, Plantae, and Animalia. There has been discussion of putting Protista either at the base of the three-way division between Fungi, Plantae, and Animalia or giving it a separate branch. A seven-kingdom system is starting to catch on, one based on biochemical evidence, with the Monera kingdom replaced by two kingdoms—**Archaebacteria** and **Eubacteria**. Archaebacteria include some interesting bacteria that live, for instance, in volcanic pools or in pools with extraordinary salt concentrations. The Eubacteria are the more familiar ones. **Slime molds** are put into a separate kingdom, and Protista are grouped separately. Another interesting feature of this seven-kingdom system is that it shows how mitochondria, flagella, and chloroplasts were acquired by eukaryotic cells from the Eubacteria. These organizational schemes are not permanent and simply reflect the best knowledge of the scientists of that time mixed with a bit of inertia.

## Introduction

The levels of organic life are atoms, molecules, macromolecules, cells, tissues, organs, systems, individuals, and societies. Anatomy deals with the levels of tissues, organs, and systems. Each organelle of a single-celled organism does a job that contributes to the total functioning of the cell. In **multicellular** organisms, each cell has become more specialized and is part of a tissue. The tissues are organized into organs and the organs into systems. The individual, whether it be **monocellular** or multicellular, must have structure and be able to control that structure. The individual must be able to take in, digest, distribute, and let out materials. The individual must be able to reproduce and, finally, to defend itself.

## Skeleton ✳

The first animals to appear had soft, **cartilaginous skeletons**. Cartilage is a tough, flexible material that is adequate in a watery medium but is unsuitable for living on land. The skeleton of human embryos begins as cartilage but then changes to bone during fetal development and into early adulthood when growth stops. This process is called **ossification**. The tips of bones remain covered by cartilage, which acts as a slippery cushion at the joints. Cells called **osteocytes** invade the cartilage and secrete **collagen**, a fibrous protein, and **calcium phosphate**, a hard mineral with needle-like crystals. The combination of tough fibers and hard mineral make **bones** hard and inflexible.

The unit of bone structure is the **Haversian system,** a small, round structure filled with concentric circles. Bone is made of hundreds of these running side by side along its length. The center of the Haversian system is a **central canal**. Scattered between the concentric layers are

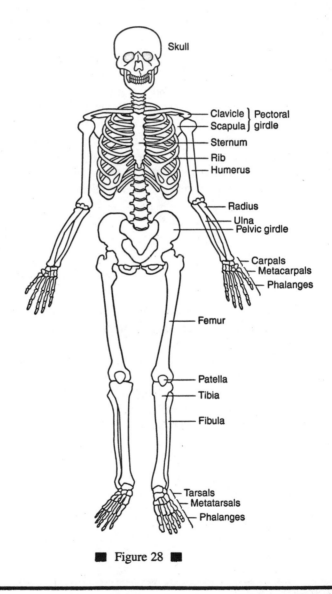

Skull

Clavicle } Pectoral
Scapula } girdle

Sternum

Rib

Humerus

Radius

Ulna

Pelvic girdle

Carpals

Metacarpals

Phalanges

Femur

Patella

Tibia

Fibula

Tarsals

Metatarsals

Phalanges

■ Figure 28 ■

spaces called **lacunae** that contain the osteocytes. The osteocytes are fed by blood vessels in the canal. The cells in bone are alive and must be nourished. A characteristic of **connective tissue** is that each cell is surrounded by a matrix that supports and anchors the cells; therefore, bone is a connective tissue.

**Compact bone** is found along the length of a bone and is dense and strong. The ends of a bone are composed of **spongy bone**. Through the center of the long bones is a large canal containing **marrow.** Yellow marrow is full of stored fat. Red marrow is found at the ends of long bones and is involved with red blood cell and blood platelet production. Bones are sheathed with a fibrous coating called **periosteum.**

**Joints** are held together by **ligaments** and cushioned by pads of cartilage. There are five kinds of joints. **Pivot joints** are like the joint between the skull and the top vertebrate. The elbow is a **hinge joint. Fixed joints** are found in the **sutures** of the skull where several plates seem sewed together. **Gliding** joints are seen in the wrists, and the shoulder joint is an example of a **ball and socket joint.**

## Muscles

There are three kinds of muscle: (1) **striated,** or voluntary, muscle, which is attached to bone; (2) **smooth,** or involuntary, muscle, which is connected to tissue; and (3) **cardiac** muscle, which appears to be a combination of striated and smooth muscle and is found in the heart. Muscles can only contract and so are always paired with opposite (**antagonistic**) muscles. For example, the **biceps** bends your arm, and the **triceps** straightens your arm. The **origin of the muscle** is usually closer to the midline of the body, and the **insertion of the muscle** is on the bone that moves. Between origin and insertion is the **belly of the muscle.** Muscles are attached to bones by **tendons.**

The microscopic examination of striated muscles shows definite bands and zones in the muscle fibers. Muscle fibers (the muscle cells) are filled with **myofibrils,** which are composed of segments called

**sarcomeres,** the contractile unit of the muscle. One sarcomere is connected to the next at the **Z line.** Thin **actin** filaments extend from the Z lines almost to the center of the sarcomere from each side. Thick **myosin** filaments float between the actin filaments. Oar-like extensions of the myosin, called **cross bridges,** connect to the actin.

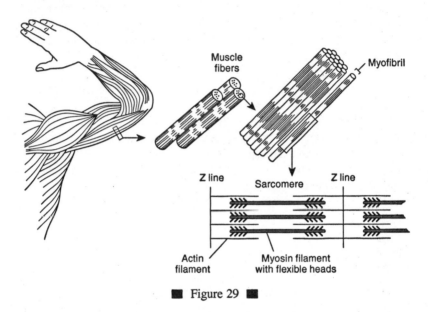

Muscle fibers

Myofibril

Z line

Sarcomere

Z line

Actin filament

Myosin filament with flexible heads

■ Figure 29 ■

When a nerve impulse strikes the muscle fiber, it penetrates and causes **calcium ions** to flood the sarcomere. The calcium ions allow ATP to expend energy to make the cross bridges work like oars pulling the Z lines closer. When the nerve impulse has passed and is not renewed, the calcium ions are pumped out, and the contraction ceases. The mitochondria use up a great deal of oxygen producing ATP for this purpose. The muscle cells store an initial supply of oxygen on molecules of **myoglobin.** When this supply runs out, oxygen is supplied from the blood. Fermentation occurs in the mitochondria when not enough oxygen can be supplied, producing very little ATP and lots of **lactic acid.** The muscle is sore for some time because of the lactic acid.

## Endocrine System ✳

The endocrine system is composed of several **ductless glands** scattered around the body. These glands secrete **hormones** (chemical messengers) into the blood. Hormones cause specific responses in **target cells**. The target cells have receptor sites in the proteins protruding from their surfaces. Interaction with the receptor site causes a conformational change in the membrane protein and sends a second messenger from the inner surface of the membrane to the nucleus. Some hormones may actually penetrate the cell membrane.

The production of hormones is a carefully controlled process. The glands of the endocrine system monitor the body's activities and interact in complex ways to regulate the internal environment. A common interaction is called **negative feedback inhibition**. An example is found in the regulation of **thyroxin (TH)** levels. Thyroxin, a hormone produced by the thyroid gland, accelerates the rate of cell respiration. The hypothalamus sends out **thyrotropic-releasing hormone (TRH)** when thyroxin levels drop. The TRH stimulates the anterior pituitary to produce **thyroid-stimulating hormone (TSH)**, which stimulates the thyroid to produce TH. The increased concentrations of TH inhibit the hypothalamus and anterior pituitary from producing TRH and TSH.

**Hypothalamus.** The **hypothalamus** is the base of the brain. It secretes a variety of hormones called **releasing hormones**, which are secreted into a capillary that is connected directly to another capillary bed in the anterior pituitary. Sensory cells communicate with the brain and the hypothalamus, which passes the message to the pituitary. Sensory deprivation causes breakdowns in hormone balance.

**Pituitary.** The **pituitary**, composed of an anterior and a posterior lobe, is connected to the hypothalamus by a stalk. The posterior lobe secretes **oxytocin**, which causes uterine contractions, and **vasopressin**, which constricts the arterioles causing blood pressure to rise and urine production to slow down. The anterior lobe is more versatile. It produces hormones that stimulate milk production (**prolactin**), affect growth (**somatotropic hormone, STH**), stimulate the thyroid (TSH), and stimulate the adrenal gland (**adrenocorticotropic hormone, ACTH**). It also produces two **gonadotropic hormones—follicle-stimulating hormone (FSH)** and **luteinizing hormone (LH)**—which help regulate gonad activity.

**Thyroid.** The **thyroid** gland produces thyroxin, which stimulates metabolic activity. It also produces **calcitonin**, which keeps the blood calcium levels low. The control of the thyroid has already been mentioned.

**Parathyroids.** The **parathyroids** sit on the thyroid. They produce **parathyroid hormone (PTH)**, which increases calcium levels in the blood by increasing the uptake of calcium from intestines, slowing the calcium excretion, and increasing the release of calcium from bones.

**Adrenal glands.** The **adrenal glands** have a **medulla** in the center covered by a **cortex**, two parts which have separate origins and different functions. The medulla secretes **adrenalin**, which stimulates the body for "fight or flight" reactions. It also secretes **noradrenalin**, which calms the body down. The cortex produces whole groups of hormones called **steroids**, which are derived from cholesterol and which control the conversion of fat to carbohydrate (**glucocorticoids**), the stimulation of the kidneys to excrete less salt (**mineralocorticoids**), and the stimulation of secondary sex characteristics (**cortical sex hormones**).

**Pancreas.** The **pancreas** is a composite of **exocrine cells** that secrete through ducts into the intestine and **endocrine (ductless) islet cells** that secrete into the blood stream. **Beta islet cells** produce **insulin**, which reduces glucose concentrations in the blood by stimulating muscle cells to absorb glucose and by speeding up reactions in the liver that convert glucose to glycogen while inhibiting the breakdown of glycogen. The **alpha islet cells** produce **glucagon,** which does the opposite of insulin, thus helping to balance the glucose levels.

**Pineal gland.** The **pineal gland** is a lobe in the rear portion of the forebrain. In lower animals, it was light sensitive and controlled **circadian rhythms** by the production of **melatonin.** In higher animals, it is no longer a light-sensing organ itself and may now help communicate neural information about light to the hypothalamus, which controls the anterior pituitary's production of gonadotropins.

**Testes.** The interstitial cells of the **testes,** when stimulated by LH from the anterior pituitary, produce **testosterone,** which, when combined with FSH from the anterior pituitary, stimulates sperm production. Testosterone helps maintain secondary male characteristics. **Androgens** other than testosterone are produced in the adrenal cortex.

**Ovaries.** Stimulated by FSH from the anterior pituitary, **ovaries** produce **estrogen,** which stimulates the growth of the lining of the uterus and the development of secondary sexual characteristics in the female. When stimulated by LH, the ovaries produce **progesterone,** which prepares the uterus for implantation and helps maintain pregnancy.

**Prostaglandins.** **Prostaglandins** are produced by cell membranes in most cells of the body. Unlike other hormones, they are fatty acids, effective at very short range and potent in extremely small concentrations. They are immediately broken down in the body. Prostaglandins increase sensitivity to pain, cause contractions of smooth muscle, and play a part in the immune system. **Aspirin** works by blocking the production of prostaglandins.

## Nervous System  ✳

**Nerve cells (neurons)** may have evolved from endocrine cells. Neurons also secrete hormones, which in this case are called **neurohormones**. One major difference is that an extension of the neuron extends to the surface of the target organ. The degree of control possible with a nervous system is much quicker and more precise than the control possible with an endocrine system.

**Nerve cell anatomy.** In general, **dendrites** carry impulses toward the cell body and **axons** carry impulses away. The **cell body** contains the nucleus. Dendrites are often more highly branched and shorter than axons. Synaptic terminals are at the ends of the axon and transmit signals to the target dendrite. **Neuroglia cells** (often called **glia cells**) surround the neurons in the **central nervous system (CNS)**—the brain and the spinal cord. Glia cells help maintain a steady environment for the neurons and, in some cases, act as conduits for growing nerve cells. **Myelin sheaths** derived from glia cells or from **Schwann cells** help to insulate a long axon from its surroundings. The **nodes** between the sheath cells are at roughly even intervals and actually speed up nerve impulses by allowing the impulse to jump ahead. The impulse in a myelinated axon approaches 200 meters per second.

The **peripheral nervous system** is made up of those neurons whose axons extend out of the CNS. The cell bodies of peripheral nerves may bunch up in a **ganglia**. Bundles of axons form a **nerve**. Nerves from the body approach both sides of the spinal cord toward the gap between each vertebra and divide into dorsal and ventral branches. The dorsal is composed of **sensory neurons** and goes through a swelling called the **dorsal root ganglia**, containing the sensory neuron cell bodies, before entering the **spinal nerve cord**. The lower branch is the **ventral root** and is composed of **motor neurons** carrying instructions to muscles. The sensory nerves and the motor nerves join just outside the spinal cord making a nerve with signals going both ways. A cross section of the spinal nerve cord shows a darker, butterfly-shaped area surrounded by a whiter area. The whitish area is composed of nerves running up and down the spinal cord. The fat-filled myelinated sheaths give it its color. The darker central area is made up of the cell bodies of **interconnecting neurons**.

**Reflex arc.** The simplest example of nerve control is the **reflex arc**. Imagine that a person touches something sharp with her finger. The pain sensory neurons are stimulated, and an impulse is sent toward the cell body in the dorsal root ganglia and continues along the sensory axons into the spinal cord. In the spinal cord, the impulse is routed to the appropriate motor neuron via an interconnecting neuron. The impulse zips out the ventral root causing the correct muscle to contract and pulling the hand out of danger. This series of events is the reflex arc. Other interconnecting neurons in the gray matter of the spinal cord send the signal to the awareness centers in the brain that evaluate the problem and send instructions to remedy the situation. While the awareness of the pain is growing, the actual reflex arc has already caused the hurt finger to be moved out of danger.

**The autonomic nervous system.** The reflex arc described above takes place in the conscious, or **somatic**, nervous system. The **autonomic nervous system (ANS)** is not usually under conscious control but is to a great extent controlled by the hypothalamus. The ANS innervates the smooth muscles in the blood vessels and digestive tract and is involved in control of the heart, lungs, and reproductive and respiratory systems. The processing of signals in the ANS occurs in ganglia outside the CNS. The ANS is divided into two parts: the **sympathetic** and **parasympathetic systems**. The sympathetic system mimics the effects of adrenalin in terms of stimulating the body for action; the parasympathetic system quiets the body. Almost every internal organ is hooked to both of these systems.

**The nerve impulse.** The **nerve impulse** is not an electrical pulse; it is a wave of opening and closing gates that allows the flow of ions. The cell membrane in a resting neuron is **polarized**—positive outside. The polarity is maintained by an active sodium-potassium pump. An **action potential** initiates an impulse by causing **gated channels** in the membrane to open, allowing a squirt of sodium atoms to diffuse into the cell. This action causes a local reversal of polarization. One gate affects the next as this reversal of polarization travels like a wave along the neuron. A fraction of a second after a gate opens it becomes permeable to potassium ions, which flood outward to re-establish the original polarization and make the neuron ready for the next action potential to start another impulse. The impulse involves only those few ions close to the inner and outer surfaces. Even if the sodium-potassium pump were to stop, nerve impulses could be generated for a considerable time.

**Synapses.** The **synaptic terminal,** or bouton, is a swelling at the end of an axon. The axon branches at its end, so there may be many boutons. Each bouton lies close to the target cell membrane. A **synaptic cleft**, or gap, between the **pre-** and **postsynaptic membranes** is around twenty nm across. The bouton is full of **presynaptic vesicles,** which contain **neurotransmitters.** A depolarization impulse sweeping around the bouton causes the vesicles near the membrane to release their neurotransmitters into the synaptic cleft. These chemicals diffuse across the cleft to receptor molecules in the postsynaptic membrane. **Acetylcholine** is the neurotransmitter in vertebrate muscles. Immediately after its release, **acetylcholinesterase** quickly destroys the acetylcholine so that new signals can be sent across the gap.

Other known neurotransmitters are **serotonin, dopamine**, and **noradrenalin** (last seen as a product of the adrenal cortex). These neurotransmitters may stimulate or inhibit. The **presynaptic neuron** is often in contact with boutons from other nerve cells. These other nerve cells are **facilitory interneurons** or **inhibitory interneurons.** Their action is to sensitize the presynaptic neuron to an impulse or to stop the impulse.

**Summation** occurs when many excitory stimulations are necessary to trigger a postsynaptic impulse. Some psychic disorders have been traced to the action of neurotransmitters. Severe depression, for example, is the result of too little serotonin.

## Sensory Receptors

**Sensory organs** react to their environment. The impulses that are sent to the brain are called **sensations**. The brain's interpretation of these impulses is called a **perception**. **Sensory receptors** are modified nerve cells that have specialized to react to some form of energy—heat, light, pressure, and chemicals. **Exteroreceptors** receive information from outside: sights, sounds, smells, touch, and taste. **Interoreceptors** receive information from the inside of the body: muscle stretching, blood pressure, concentrations of oxygen and carbon dioxide, and body temperature. Receptors can be categorized further into mechanoreceptors, chemoreceptors, electromagnetic receptors, thermoreceptors, and pain receptors.

**Mechanoreceptors** are stimulated by mechanical energy: sound, touch, stretching, and motion. In humans, **Pacinian corpuscles** deep in the skin respond to heavy pressure. Lighter touches are picked up by two types of receptors closer to the surface called **Meissner's corpuscles** and **Merkel's disks**. Stretch receptors detect the stretching of muscles. The knee-jerk reflex begins as the leg muscle is stretched. **Hair cells** are cells with cilia-like projections that are sensitive to motion caused by fluid movements, as in the inner ear, or sound waves. Real hair often has nerve webs around the base; these webs detect any motion of the hair.

**Chemoreceptors** detect concentrations and identities of molecules in vapor or solutions—**taste** and **smell** as well as internal sensors of blood concentrations of various chemicals. **Electromagnetic receptors** are sensitive to light (**photoreceptors**). **Thermoreceptors** help maintain body temperature and are sensitive to heat (**Ruffini's end organs**) and cold (**end bulbs of Krause**). **Pain receptors** are thought to be naked nerve endings called **nociceptors**.

**The eye.** A clear **cornea** at the front with the help of a **lens** focuses an image on the **retina**. The amount of light is regulated by an **iris**, which in humans is colored. The **pupil** is the opening in the iris that lets in light according to its diameter. A clear, jelly-like material between the cornea and the lens is called **aqueous humor**. Filling the large space behind the lens is a similar fluid called **vitreous humor**. The cornea is continuous with a nontransparent, tough coating called the **sclera**, which covers the rest of the eye. Under the sclera is the **choroid layer**. The lens is held in place by **suspensory ligaments** and wrapped by **ciliary muscles**. For close viewing, the ciliary muscle contracts, making the lens thicker and shortening its focal length, an adjustment called **accommodation**. Blood vessels enter the eye next to the **optic nerve**. The pattern of blood vessels visible on the retina is as distinctive as a fingerprint.

The **retina** is a flimsy layer covering the inside of the eye. It contains **rods** (photorceptors sensitive to black and white light) and **cones** (photoreceptors that can distinguish colors). The cones are most concentrated in the **fovea**, the center of focus for images; farther from the fovea, the percentage of rods increases. Where the optic nerve enters the back of the eye is the **blind spot** because there are no photoreceptors there.

Rods and cones are distinct from each other, but both contain stacks of disks whose membranes contain **opsin** and molecules of **retinal**, a light-sensitive chemical synthesized from vitamin A. On being struck by light, the retinal isomerizes, changing the polarity of the membrane, which starts a signal toward the brain. In rods, the main opsin is **rhodopsin**. Light isomerizes a part of its structure to change the level at which it floats in the membrane, with the new level apparently making the rod even more sensitive to low levels of light.

**The ear.** The ear is responsible for two senses: **hearing** and **balance**. Most people use the word *ear* to describe just the **pinna**, or outer ear. An **auditory canal** leads to the **tympanum** (the eardrum), which is pushed in and out by the pressure waves of sound. These vibrations are transmitted by three bones—the **ossicles**, called the malleus, incus, and stapes—to the **oval window** on the surface of the **vestibule**, which is the "mouth" of the **cochlea**. Near the oval window is the **round window,** which pushes out when the incoming vibration pushes the oval window inward. The cochlea is a long tube with one end at the oval window and the other at the round window. This tube is folded at its tip half-way along its length so that the second half returns along the first half. The half associated with the oval window is the **vestibular canal,** and the half associated with the round window is the **tympanic canal.** Between these canals is another tube called the **cochlear duct.** The fluid in the vestibular and tympanic canals, which are continuous, is called **perilymph.** The fluid in the cochlear canal is called **endolymph.** These three tubes are coiled to make the whole cochlea.

The cochlea is in the inner ear. The middle ear contains the ossicles. The **auditory tube (Eustachian tube)** leads from the middle ear to the pharynx. Pressure differences between each side of the tympanum are equalized by air entering or leaving the middle ear through the auditory tube. This creates the familiar popping sound that you hear when you change altitude suddenly.

The actual hearing occurs in the **organ of Corti**, which sits on its **basilar membrane** on the tympanic canal and extends into the cochlear duct. The organ of Corti has hair cells with their cilia resting on the nearby **tectorial membrane**. When pressure waves in the tympanic canal bounce the basilar membrane, the hair cells are shoved against the tectorial membrane. Bending the cilia of the hair cells causes an action potential, and a signal is sent to the brain via the **auditory nerve**. The origin of the signal along the length of the basilar membrane is related to the wavelength of the sound. The brain distinguishes pitch by sensing the portion of the basilar membrane that sent the signal.

Balance is a complex sense. Part of our information about balance comes from our eyes, another part from the stretch receptors in our muscles, and a third part from the motion sensors and attitude sensors in the inner ear. Anyone who has ever been dizzy will attest that these three inputs do not always agree. The vestibule of the inner ear has two chambers called the **saccule** and the **utricle**. Hair cells project their cilia into the gelatinous liquid in the saccule and utricle. Tiny particles of calcium carbonate called **otoliths** are on the cilia. Tilting your head one way or another causes the hair cells to bend as they are tilted. The action potentials created by this bending send signals to the brain regarding the direction of up and down.

The utricle leads into three loops mutually perpendicular to each other called the **semicircular canals**. Hair cells extend into their fluid. When you turn your head, inertia causes this fluid to remain stationary as the hair cells are swept through it and bent. The signal that they send tells the brain what direction your head moved. Water in a soup bowl does the same thing. If the bowl is rotated, the water does not rotate with it at first. After a short time, friction causes the fluid to rotate with the bowl. When the bowl is stopped, the fluid continues to move. Dizzy people perceive their surroundings to be still moving even after they have stopped moving.

## Digestion ✕

**The mouth.** **Digestion** is the process of breaking down food materials and absorbing them into the body. This process occurs along the **alimentary canal** with the help of accessory glands. The **teeth** break the food into smaller chunks, thus increasing its surface area. **Salivary glands** around the mouth secrete saliva to moisten the food. Saliva contains **mucin,** a slippery glycoprotein; buffers to neutralize any acids or bases; bacteriocides to kill unwanted organisms; and **salivary amylase,** which breaks starches into disaccharides. The **lips, cheeks,** and especially the **tongue** move the food around while it is being **masticated** (chewed). The tongue rolls the chewed food into a **bolus** and moves it to the back of the mouth where it is swallowed.

The bolus enters the **pharynx** (throat) where it has to cross over the **trachea** (windpipe) to get to the **esophagus**, a muscular tube leading to the stomach. Swallowing causes the **epiglottis** (a cartilaginous flap) to cover the trachea. In the esophagus, circular muscles above the slippery bolus constrict, squeezing the bolus downward. Ahead, the circular muscles relax allowing the esophagus to become wider, and the longitudinal muscles constrict to shorten the esophagus. This action continues in a wave down the esophagus, called **peristalsis**. Striated muscles at the top are under voluntary (conscious) control, but very soon smooth muscles (involuntary) take over. The rest of the alimentary canal has only smooth muscle.

**The stomach**. The esophagus enters the **stomach** just below its top. The lining of the stomach is a thick **mucosa** made of a layer of mucus secreting epithelial cells. Numerous **gastric pits** lined with epithelial cells are found in the mucosa. Gastric pits also contain **chief cells**, which secrete **pepsinogen,** and **parietal cells**, which secrete **hydrochloric acid**. The acid causes pepsinogen to become **pepsin,** which when activated can activate other pepsinogens and, most important, break protein down into short polypeptide segments. Digestive enzymes which are secreted in an inactive state are called **zymogens.**

The lining of the stomach is protected by its mucus covering and by the fact that acid and pepsinogen are usually secreted only when food is present. A break in the mucus coating leads to an ulcer where the stomach lining begins to be digested. Behind the mucosa are strong muscles that can churn the food and slosh it around. This mixture is called **chyme**. This churning action of the smooth muscles of the stomach produces hunger pains and growling when the stomach is empty. "Heartburn" is the result of a bit of acid chyme being squeezed back into the esophagus.

Some cells in the stomach wall act as endocrine glands. When triggered by signals from the brain or by the presence of food, these cells secrete **gastrin** into the blood. Gastrin causes the parietal and chief cells to secrete their enzymes. The **pyloric sphincter** keeps the exit of the stomach shut until the chyme is properly mixed.

**The small intestine**. At the correct moments, small amounts of chyme are released into the **small intestine**. Contact of this material with the wall of the small intestine triggers the release of **enterogastrin** into the blood from specialized cells in the wall of the small intestine. The enterogastrin inhibits the production of gastrin and begins to shut down the secretion of **gastric juices** in the stomach. The acidic chyme is neutralized at the beginning of the small intestine so that other enzymes can go to work.

The **pancreas** makes several enzymes that are sent down the **pancreatic duct** to the small intestine. **Pancreatic amylase** continues the breakdown of polysaccharides to disaccharides. Disaccharidases, mainly **maltase**, break the disaccharides into monosaccharides, mainly **glucose**. These processes occur on the lining of the small intestine. **Trypsin** and **chymotrypsin** are pancreatic enzymes that break polypeptides at certain places. **Carboxypeptidase** and **aminopeptidase** chew off one amino acid at a time. These enzymes are generally inactive but are activated by **enterokinase**. Fat is hard to digest; **lipase** breaks fats into **fatty acids**; **bile** from the **liver** helps to **emulsify** the fatty acids.

Most of the digestion has been completed by the time the chyme has passed through the **duodenum,** the first section of the small intestine. The rest of the small intestine, the **jejunum** and the **ileum,** is specialized for absorption. Their surfaces are convoluted and the convolutions covered with **villi,** tiny finger-like projections. The absorptive cells of the villi are covered with **microvilli.** All of this increases the surface area greatly. Each villi has a network of capillaries inside and contains a projection of the lymphatic system called a **lacteal.** Carbohydrates and amino acids are picked up by the blood, while most fatty acids go into the lacteal. Some of these materials pass through the membranes of the absorptive cells passively, but most are actively transported.

All the blood vessels leaving the small intestines are collected into a large vessel called the **hepatic portal vein** and sent through the liver. In the liver, the portal vein divides into a fine network of capillaries. The job of the liver is to regulate the concentrations of materials in the blood. If there is too much glucose, the liver converts some of it to glycogen. If the amino acid concentration is too low, the liver causes more amino acid to be released into the blood. The blood entering the liver may have a wide variety of concentrations of materials from the intestines, but the blood leaving the liver has fairly constant levels of dissolved nutrients.

**The large intestine.** The junction of the small and large intestines has a Y shape. The third arm of the Y is called a **cecum**. In humans, the cecum is small and ends in a projection called the **appendix**, a vestigial organ with no function. This juncture is low on the right side of a person's abdomen. The **ascending colon** travels upward from this point. Just under the stomach, the **large intestine** turns to go across the abdominal cavity. This is the **transverse colon**. Then, it turns and descends, becoming the **descending colon**. The large intestine finishes the job of absorbing water to leave a semisolid residue (**feces**), which is held in the **rectum** until it may be voided (**defecation**).

The large intestine is where a symbiotic organism called *E. coli* lives. These bacteria provide us with vitamin K. They and other bacteria digest bits of food material that were not digested or not absorbed, often producing intestinal gas. Nearly half of human feces is composed of these intestinal guests. *E. coli* counts detected in streams, lakes, and oceans provide an idea of the level of human sewage in the water.

## Respiration ✕

**Respiration** involves taking in oxygen and expelling carbon dioxide. In humans and other vertebrates, the **lungs** are the major organ for this process. Breathing through the nose allows the air to come closer to body temperature, to acquire moisture, and to be somewhat filtered. The air then passes through the trachea, which splits into two **bronchial tubes**. These divide into finer and finer tubes (**bronchioles**) until a cluster of little bags called **alveoli** is reached. The alveoli are thin-walled sacs whose surface area is close in size to the area of a tennis court and lined with capillaries. The surface of each alveolus is coated with a thin layer of moisture through which oxygen and carbon dioxide pass. The **pulmonary artery** carries oxygen-poor blood to the lungs, and the **pulmonary vein** carries oxygen-rich blood back to the heart. Oxygen in the alveoli is exchanged for carbon dioxide in the blood.

The lungs are in the **thoracic cavity**. The **pleural membrane,** which lines the cavity, is smooth and secretes a small amount of lubricating fluid. The space between the lung and the pleura is closed to the outside. Therefore, when the thoracic cavity expands, the lungs must expand also. Muscles pull the ribs more perpendicular to the backbone, and the **diaphragm**—a domed sheet of muscle between the lungs and the stomach—flattens out to increase the volume of the thoracic cavity. This is called **negative pressure breathing**.

The **partial pressure of a gas** is a measure of its concentration in air or liquid. Gas will diffuse from a greater partial pressure to a smaller partial pressure. In the body, partial pressure increases for carbon dioxide and decreases for oxygen during normal metabolism. The fresh air in the alveoli has more oxygen and less carbon dioxide so that oxygen diffuses into the blood and carbon dioxide diffuses into the air in the alveoli. The partial pressure of oxygen in the air diminishes quickly at higher altitudes. People sitting still show symptoms of oxygen deprivation above 12,000 feet and must use an oxygen mask if they are piloting an airplane. Athletes notice the difference of only a few thousand feet in altitude.

A small percent of carbon dioxide dissolves in the blood, but most of it becomes carbonic acid, which gives off a hydrogen ion to become a bicarbonate ion in **red blood cells**. The hydrogen ion is accepted by protein in the red blood cell. The red blood cell acts as a buffer in this manner so that the pH level of the blood stays constant. The process is reversed at the alveoli. **Hemoglobin** molecules cover the surface of red blood cells. Hemoglobin can hold an oxygen molecule so that not much is actually dissolved in the blood.

## Circulation ✕

**Blood vessels.** The system for pumping blood around is called the **cardiovascular system**. A 160-pound person has about 6 liters of blood. **Arteries** lead away from the heart, are under the most pressure, and have muscular walls. Arteries become **arterioles,** which are the same but smaller and with less pressure. The arterioles have sphincter muscles that control the amount of blood flowing through. Arterioles become **capillaries,** which are barely big enough for a red blood cell to get through. Blood moves fastest in the arteries and slowest in the capillaries because the total cross-sectional area is smallest in arteries and greatest in capillaries. Gas and nutrient exchange occur at the capillary level. The blood continues on to **venules** and then to **veins** and finally back to the heart. The return-flow vessels have much thinner walls and the lowest pressure. Another difference is that veins have **valves,** which prevent back flow of blood.

**The blood. Plasma** is a straw-colored liquid that makes up about 60% of the blood. Plasma is water containing all of the chemical salts (**electrolytes**) and all of the nutrients and wastes of the cells. Additional plasma proteins (**albumins**) make the plasma slightly hypertonic to the cells, preventing loss of water from the blood stream to the cells. These proteins also carry fat and other insoluble molecules. **Fibrinogens,** another soluble plasma protein, play a chief role in blood clotting. **Globulins,** which include molecules of the immune system, are a third group of plasma proteins.

Red blood cells (**erythrocytes**) are loaded with hemoglobin to carry oxygen. These are cells without nuclei, which live about 120 days and are replaced by new cells made in bone marrow. Amoeba-like white blood cells (**leukocytes**) are larger, have a nucleus, and are not as common as erythrocytes. They consume damaged tissue and invading bacteria. Not confined to the blood vessels, they can squeeze through capillary walls into the interstitial fluids. Another blood component is **platelets,** which are produced in bone marrow and are critical to the clotting process.

Blood clotting occurs when a protein known as **tissue factor** is exposed to the blood, which occurs when the vessel wall is broken or when platelets are broken. The process occurs as follows: **prothrombin** is converted to **thrombin** by the catalytic action of **thromboplastin.** The thrombin then converts **fibrinogen** to **fibrin.** The fibrin sticks to everything in the area and forms a net that traps red blood cells and generally clogs up the area. This clog is called a **scab**, and it seals off the wound.

**The heart.** The human heart has four **chambers**, two on top and two under. The diagram of a heart faces you, so its right side is on your left. Blood enters the chamber called the **right atrium** on top from the **superior** and **inferior vena cava**. When the right atrium chamber contracts, blood is pushed through a one-way valve called the **right atrioventricular valve** into the **right ventricle**. From this second chamber, the blood is pumped through another one-way valve, the **right semilunar valve**, into the **pulmonary arteries** on its way to the lungs. All of this blood returns to the heart from the lungs via the **pulmonary veins** and into the **left atrium**. From this third chamber, blood is pumped through the **left atrioventricular valve** into the **left ventricle**. The left ventricle is the most muscular chamber of all and exerts the most pressure, sending the blood up and out through the **left semilunar valve** into the **aortic arch**. The aortic arch branches into the **carotid arteries** to the head and the **brachial arteries** to the arms and, as the **aorta**, bends downward toward the legs.

The walls of arteries are muscular and elastic. When the left ventricle contracts, the aortic arch fills up like a long balloon under pressure. The pressure produced by the ventricle is called the **systolic pressure**. When the left ventricle relaxes, the semilunar valve slams shut, and the blood is squeezed by the artery. The pressure caused by the elasticity of the major arteries is called the **diastolic pressure**.

The heart responds to several control mechanisms. First, it has a built-in beat initiator called the **sinoatrial node**. This group of cells has nerve and muscle cell characteristics and is at the juncture where the veins join the right atrium. They contract periodically on their own, sending out an impulse that passes over the atria like a wave and causing them to contract. This wave also reaches a similar group of cells between the right atrium and ventricle called the **atrioventricular node**. This node contracts and sends off a signal that travels in a wave through the muscle cells in the ventricles. Heart action, then, consists of a contraction of the atria followed quickly by a contraction of the ventricles. The *lub dub* sound is the closing of the atrioventricular and then the semilunar valves. First, the atrioventricular valves snap shut as the ventricles begin to bear down on the blood within them, and then the semilunar valves slam shut as the ventricle relaxes and the arteries try to squeeze the blood back into the heart.

**The lymphatic system.** The arteriole end of a capillary has enough **hydrostatic pressure** to cause it to leak fluid out faster than the osmotic pressure can pull fluid in. At the venule of the capillary, the hydrostatic pressure is less, and the net flow of fluid is into the capillary. However, more fluid is lost at the front of a capillary than is reclaimed at the end of a capillary. The result is a fluid build-up in the interstitial spaces. This fluid is basically plasma but is now called **lymph**. There is a specialized system of vessels called the **lymphatic system** that channels this fluid back into the blood stream. Interstitial fluid seeps into the smallest blind ends of the lymph vessels. These vessels are much like veins with thin walls and one-way valves. There is no pump. Apparently the contraction of skeletal muscles squeezes the lymph vessels, inducing the lymph to move toward the heart. There are many **lymph nodes** along the return path that act as traps for bacteria and any other antigenic materials. Finally, the lymph empties into the superior vena cava just outside the heart.

## Excretion and Water Balance ✕

**The kidneys.** The waste in the blood is removed by the **kidneys**. The aorta sends out branches left and right called the **renal arteries**, which direct blood under fairly high pressure into the kidneys. The artery quickly branches into capillaries, which form hundreds of thousands of clusters in the **cortex** of the kidney called **glomeruli**. The functional unit of the kidney is the **nephron**. On the front end of each nephron sits a structure called **Bowman's capsule**, which surrounds each glomerulus. A great deal of fluid leaks out of the glomerulus and is

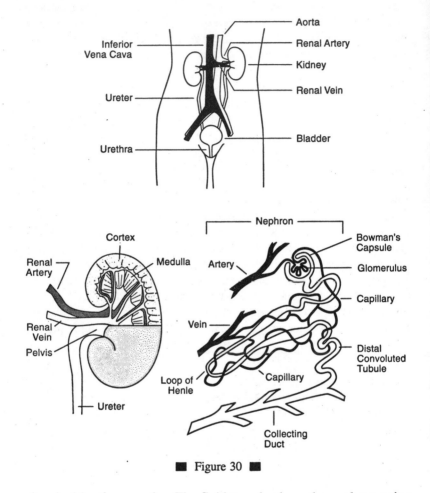

■ Figure 30 ■

absorbed by the capsule. The fluid travels along the nephron going through the **proximal convoluted tubule** and then the **loop of Henle,** which descends into the **medulla** and makes a hairpin turn to ascend back to its starting level. Then, the fluid passes through the **distal**

**convoluted tubule** and into a **collecting duct**. Joined by other distal convoluted tubules, the collecting duct empties into the **renal pelvis**, which sends its fluid along the **ureter** to the **bladder**. Meanwhile, the blood from the glomeruli collects and enters another capillary web around the loop before rejoining the **renal vein**.

The process of collecting waste molecules and returning the filtered fluid back to the blood begins in the glomerulus. Because of the high pressure, a lot of fluid escapes the capillary with various small molecules. All this fluid and its contents finds its way into the capsule. Larger molecules like plasma proteins are not pushed out of the capillary. The fluid in the capsule (the **filtrate**) is similar to plasma as far as its small molecules go. In the proximal convoluted tubule, selective secretion takes place as unwanted materials in the interstitial spaces are actively or passively transported into the filtrate. At this stage and later, **reabsorption** allows needed materials like sugar, vitamins, and minerals to be returned to the blood.

Quite a bit of water is lost from the filtrate in the descending limb of the loop of Henle because the medulla is high in salt and therefore hyperosmotic. In the ascending limb, salt is actively pumped out of the tube both to maintain the salt gradient in the medulla and to bring the concentrated fluid in the ascending limb into osmotic balance. The filtrate now is passed through the distal convoluted tubule to a collecting tubule where it again is conducted to the medulla. More water passively leaves the filtrate and goes into the hyperosmotic fluid of the medulla. The fluid also loses some **urea**, a waste product, into the medulla, further increasing its hyperosmotic properties. The filtrate is now **urine** and ready to be expelled from the body.

**Regulation of the kidneys. Antidiuretic hormone (ADH)** is produced by the hypothalamus and stored in the pituitary. **Osmoreceptor cells** in the hypothalamus monitor the osmotic balance of the blood and trigger the release of ADH when the osmotic concentration of the blood rises. The ADH causes more water to be reabsorbed through the distal convoluted tubule and the collecting duct. As the blood becomes more dilute, it causes the hypothalamus to stop releasing ADH. **Diuretics**

reverse this, increasing the amount of urine produced. The purpose of controlling the kidney is to maintain a **homeostatic** blood concentration, a concentration that does not change.

## Reproduction ✗

**Gametogenesis.** **Gametes** are either **egg cells (ova)** or **sperm cells.** The production of egg cells is called **oogenesis** and the production of sperm cells is **spermatogenesis.** Both processes involve **meiosis** to produce haploid cells—gametes which contain one set of chromosomes. When a gamete from one individual fuses with a gamete from another individual of the opposite sex, **fertilization** occurs and a **zygote,** or **fertilized egg,** is formed. The zygote contains the diploid number of chromosomes, one set from each parent.

Oogenesis goes through meiosis in the expected way, but **cytokinesis** is unequal. Oogenesis produces one large cell, the **ovum,** which contains most of the cytoplasm, and three tiny **polar bodies,** which are reabsorbed. In a female, oogenesis is almost complete by the time she is born. Her **ovaries** store all the ova that she will ever produce. The ova mature and are released for fertilization at intervals during her fertile years. A mature ovum is large, visible to the naked eye in ideal lighting. Most of its contents are food (**yolk**) for the developing embryo until it can start getting nutrients from its mother's blood.

Spermatogenesis begins when the male reaches sexual maturity. The process involves meiosis, with an even distribution of cytoplasm among the resulting gametes. These gametes go through a maturation process as they change to mature sperm cells. An **acrosome** at the tip of the sperm cell contains enzymes to help it penetrate the egg cell's membrane. Next is the tightly packed set of chromosomes, the nucleus. A long tail extends from the head. The tail is thicker near the head where many mitochondria are clustered. The rest of the tail contains the axial filament, a flagella, covered by a fibrous sheath. The sperm are formed continually in the **seminiferous tubules** of a mature male.

There are several seminiferous tubules, each wound up in a compartment of the **testes**. These tubules send sperm into the **epididymis**, where they are given time to mature and are held until an ejaculation propels them up the **vas deferens** in the **spermatic cord**. In each compartment of the testes outside the seminiferous tubules are **interstitial cells** responsible for hormone production. The two testes are held in a sac called the **scrotum**. Because spermatogenesis works best a little below body temperature, the muscles of the scrotum can relax to allow the testes to hang farther from the body and be cooler or the muscles can contract to bring the testes up against the body to raise their temperature.

**Male reproductive physiology.** When sexually excited, the **erectile tissue** of the **penis** will fill with blood and extend. Erectile tissue is also found in the clitoris in the female and in the nipples and ear lobes of males and females. **Ejaculation** occurs after a period of stimulation of the sensitive areas on the erected penis, leading to an **orgasm**. During ejaculation, the sperm are pulled out of the epididymis by peristaltic contractions of the muscular vas deferens. Both vas deferens loop up in front of the pubic bone along and beside the bladder and finally join below the bladder to become a short tube called the **ejaculatory duct**, which joins the **urethra**. The urethra carries either urine or **semen** to the tip of the penis. More than half of the semen comes from two glands on either side of the ejaculatory duct called **seminal vesicles**. During ejaculation, the seminal vesicles secrete a thick fluid containing fructose to feed the sperm and prostaglandins to stimulate the muscles of the uterus.

Another gland contributing to the semen is the **prostate gland**. It sits under the bladder. The urethra and the ejaculatory duct go through it. The prostate gland secretes a thin, alkaline fluid, which will neutralize the acidic environment of the vagina. Finally, two **bulbourethral glands**, located below the prostate gland, secrete a small amount of fluid, which may act as a lubricant. This secretion occurs before orgasm and may carry a few sperm to the tip of the penis before the later ejaculation.

**Female reproductive physiology.** The two **ovaries** produce hormones, hold **primary (immature) oocytes,** and allow the oocytes to mature and be released. Of the roughly 400,000 oocytes in both ovaries, only a fraction are ever released. The oocytes are each contained in a tough capsule called a **follicle.** The major female hormone **(estrogen)** is produced in the cells of the follicle. During the first half of the menstrual cycle, a number of follicles begin to grow. All of the follicles but one (rarely more) are somehow inhibited, and only one matures to form a blister on the side of the ovary. At the proper moment in the cycle, the egg is released **(ovulation).** The broken follicle cell thickens to form the **corpus luteum,** which secretes its hormone during the second half of the menstrual cycle. The egg is swept up by the currents generated by the cilia at the mouth of the **oviduct.** If **fertilization** occurs, it will usually happen in the oviduct. The oviduct leads to the **uterus,** a muscular organ that will hold a developing fetus. The inner lining of the uterus (the **endometrium**) has the special property of being able to become highly vascularized under hormonal influence. The bottom of the uterus narrows and protrudes into the top of the **vagina.** The opening in the bottom of the uterus is the **cervix.** The vagina reaches the outside between the urethra and the anus. This opening is protected by two thin, delicate folds of skin called the **labia minora.** And over the labia minora are the **labia majora,** two fleshy folds of skin covered by pubic hair. During sexual arousal, the labia become engorged with blood, exposing the entrance to the vagina. Also at this time, the vagina lengthens, becomes wider, and secretes a fluid that lubricates and neutralizes the interior of the vagina. During climax, rhythmic contractions occur near the opening of the vagina, and the cervix bobs into the top of the vagina. Climactic contractions occur in the uterus as well. All of this activity helps the sperm reach the oviducts.

**The menstrual cycle.** The **menstrual cycle** is controlled by the **hypothalamus.** During the **flow phase,** which lasts for the first few days of the cycle, the lining of the endometrium is shed. At this time, small amounts of the two **gonadotropins** from the anterior pituitary, **luteinizing hormone (LH)** and **follicle-stimulating hormone (FSH),**

cause several follicles to begin to grow. Usually only one follicle continues to grow and the others recede. This marks the beginning of the **follicular phase**. These two gonadotropins are secreted from the anterior pituitary when the **gonadotropin-releasing hormone (GnRH)** from the hypothalamus stimulates the anterior pituitary. The follicle which continues to grow secretes increasing amounts of **estrogen** (a **steroid**), which causes the endometrium to start to thicken again. The estrogen level increases during the follicular phase. Growing follicle cells have receptors for FSH but not LH. The estrogen reaches a concentration that kicks the hypothalamus into secreting a large dose of GnRH. The anterior pituitary responds, near the middle of the cycle, by releasing a jolt of FSH and LH. By this time, the follicle cells have receptors for LH so that LH in the blood can cause the follicle to mature and rupture. This important point in the cycle is called **ovulation**. The estrogen levels drop soon after. The LH now causes the old follicle cells to become a **corpus luteum**. The **progesterone** secreted by the corpus luteum maintains the lining of the endometrium. The cycle is now in its **luteal phase**. The lower levels of estrogen and the higher levels of progesterone act to inhibit the production of LH. Without LH, the corpus luteum degenerates and its production of progesterone falls off. At some point, the loss of LH causes the lining of the endometrium to slough off, and the cycle begins again.

**Pregnancy.** The fertilized ovum (**zygote**) implants in the endometrium a few days after ovulation and becomes an **embryo**, which immediately begins to generate hormones to alter the mother's menstrual cycle. The chief hormone is **human chorionic gonadotropin (HCG)**, which maintains the corpus luteum and thus the progesterone that is maintaining the endometrium. It is the high levels of HCG that can be detected in pregnancy tests.

The first trimester of pregnancy is the busiest for mother and child. The first division of the zygote, called **cleavage**, occurs within twenty-four hours of fertilization. In the few days that it takes the embryo to reach the endometrium, it has developed to the **blastocyst stage**. The blastocyst bores into the endometrium, and the endometrium grows around it. The **placenta** develops from one side of the blastocyst to allow nutrient/waste exchange through the vessels in the **umbilical cord**. HCG levels are high. **Morning sickness** often manifests itself. **Organogenesis** is going at a high rate as the organs form. The heart is beating by the fourth week and can be heard by the end of the third month. The **chorion** is the membrane that encloses the whole embryo and its other structures. **Chorionic villi** extend from this membrane surface into the endometrium, separating the circulation in the placenta from the mother's blood. The tiny **yolk sac** surrounds the yolk in the zygote and is quickly outgrown. The **amniotic sac** surrounds the embryo and holds fluid, which allows the embryo to float.

The second trimester is quieter and more enjoyable for most mothers. Hormone levels have stabilized, and the growing **fetus** makes a visible abdominal swelling. The third trimester is more of a strain. The fetus puts on weight. The mother's organs are crowded, causing discomfort. Birth (**parturition**) is controlled by prostaglandins from the uterus, oxytocin from the posterior pituitary, and nerve reflexes. **Contractions** begin. The cervix relaxes and **dilates**. In a normal birth, the baby is soon pushed out. Later the placenta is expelled. **Lactation** begins as **prolactin** from the posterior pituitary is no longer inhibited by progesterone. The prolactin stimulates milk production within a couple of days after birth.

## Immune System X

**Nonspecific defenses.** These are the defenses that block antigens from entering the body. The **skin** and **mucus membranes** act as a barrier. Mucus, perspiration, tears, and saliva contain **lysozymes**, enzymes which break down bacterial cell walls. The gastric juices in the stomach kill any bacteria entering with the food. The respiratory tract is protected by hairs in the nose, which act as filters, and by cilia, which push anything caught in the sticky mucus upward so that it can be expelled or swallowed.

If something penetrates the skin, the broken cells release **histamine,** which causes adjacent capillaries to dilate allowing more blood flow, swelling, reddening, and an increase in temperature. White blood cells called **phagocytes** are drawn to these broken cells and begin ingesting them and the bacteria. The most import phagocyte is the **macrophage,** an ameboid cell that oozes its way between interstitial cells, occasionally hitching a ride in the blood stream. It consumes damaged tissue and bacteria. **Lymphocytes,** another type of white blood cell, hang around in lymph nodes to catch anything that floats by. When a cell is attacked by viruses, it releases **interferon.** This class of chemicals are all small proteins. They stimulate white blood cells and cause nearby cells to resist viral attacks.

**Specific defenses.** The key to having a specific defense is to be able to recognize "self" and "nonself." **Antigens,** the invaders, are "nonself," and **antibodies,** also called **immunoglobulins,** are produced to go after them. Antibodies are produced by **B-lymphocytes** (B cells) and **T-lymphocytes** (T cells) and are proteins with a shape that allows them to fit and stick on antigens. B cells release the antibodies, and in T cells, the antibodies stick out of the membrane. An **immunocompetent** cell is one that produces receptor sites which match an antigen. That immunocompetent B cell or T cell also has the DNA to make more matching antibodies that can attack the antigen.

The immune system operates along the lines of the theory of evolution. Antibodies are not created in response to antigens. Rather, the antibody shapes are pre-evolved and then stay around uselessly until they find an antigen that they match. At that moment, the B cell or T cell to which they are attached is triggered to reproduce—to make clones of itself. **Clonal selection** is the process by which a lucky B cell or T cell with the proper antibodies makes thousands of itself to fight the antigens. After a slow **primary immune response,** the ability to meet that particular antigen is remembered in a **memory cell.** When that antigen is met again, it causes a **secondary immune response** that is faster than the first. This principle is the basis for vaccinations.

The antibodies most often seen are "Y" shaped and constructed of several protein subunits. Most of the subunits are stable and unlikely to change, but the subunits on the tops of the arms are easily mutated. It is these subunits that form a match for proteins on the antigen and stick to them. One antibody grabs two antigens, and other antibodies join in the fray, grabbing antigens. A large web of antigens held together by antibodies develops. A macrophage is able to find and eat these **precipitations**. The T cells also stick to antigens. In the case of T cells, the whole cell is involved, and the network is called an **agglutination**.

There are cases where the immune system misfunctions or stops functioning. Allergies are examples of immune system misfunction. Arthritis is a case of **autoimmunity**. Acquired Immune Deficiency Syndrome (AIDS) is a case of a virus attacking the immune system lymphocytes themselves and eventually weakening the immune system to the point that the body cannot defend itself anymore. Studies are revealing the complexity of the immune system. T cells can be **killer T cells, cytotoxic T cells, helper T cells,** or **suppressor T cells**. A macrophage cell can eat the antigen and then display the antigen on its surface, like a trophy along with **major histocompatibility complexes (MHC)**. The antigen-MHC complex causes a helper T cell to bind to it, which releases **interleukin-1** to stimulate growth and cell division of the T cell. This activation releases **interleukin-2**, which further enhances growth and division of T cells. Immunity is a complex and fascinating process.

## Kingdoms X

How many kingdoms should living things be divided into? Not so long ago, living things were either plants or animals. Today, a growing number of people using DNA matching techniques think that seven kingdoms are required: **Archaebacteria, Eubacteria, Protista, Fungi, Slime Molds, Plantae,** and **Animalia.** The majority, however, are still using five kingdoms: **Monera,** Protisa, Fungi, Plantae, and Animalia. Even with the most sophisticated techniques, the debate over the best classification groupings is continuous.

## Monera X

Monera are **bacteria.** They have no nuclear membrane and so are **prokaryotic.** They have cell walls and live in most places on the earth. They are mostly **heterotrophic,** but there are some **autotrophs** which use photosynthesis or sulfur compounds for energy in deep sea vents. The photosynthetic bacteria have developed at least three different photosynthetic systems. The photosystems, based on green pigment, may have become the chloroplasts of green plants. Bacteria often have flagella, which are the same as any flagella in the animal kingdom. Normal bacteria are called **eubacteria.** They come in a variety of shapes. **Cocci** are spherical. **Bacilli** are rod shaped. **Spirilla** are corkscrew shaped. These bacteria are solitary or combined. In pairs, cocci are called **diplococci,** in strings, **streptococci,** and in bunches, **staphylococci.** Bacilli can be in lines; then they are called **streptobacilli.**

The **archaebacteria** are the unusual bacteria that live either anaerobically in swamps, converting carbon dioxide to methane, in very salty water, or in hot springs. These strange bacteria are all chemically quite different from the eubacteria. Life has existed on the earth a little over 3.5 billion years. For almost 3 billion of those years, bacteria had the earth to themselves.

## Protista ( ? )

Protista are **eukaryotic**, usually single-celled organisms, but there are examples of **colonial** and **multicellular** protists. The members of the kingdom **Protista** are grouped by sections and by divisions. In the section **Protophyta**, the plant-like Protista, is found the division **Euglenophyta**, which includes the euglena, a single-celled organism that swims by means of a flagella. It has chloroplasts and is able to make its own food but is also able to ingest food like an animal. The division **Chrysophyta** includes the autotropic **diatoms**, which have cell walls made out of silica. Because of their vast numbers, they play an important role in aquatic food chains as photosynthetic organisms. They are a major component of **plankton**. The cells in the division **Dinoflagellata** have two flagella: one is wrapped around the cell's waist in a groove, and the other extends along the axis. They are yellow to brownish because they contain **carotenoids** and **xanthophylls**. Some dinoflagellates are able to produce light that shows up at night in calm ocean water when the water is disturbed. Seeing the waves fluorescing on the shore at night is a memorable experience. **Red tide** is a poisonous type of dinoflagellate, which blooms occasionally, coloring the water. It can kill fish, make eating filter feeders unsafe, and causes skin and eye irritation in human swimmers.

The section **Gymnomycota** contains the **slime molds.** These organisms exist for part of their life cycle as independent amoeboid cells that live in rotting logs. When the time comes to reproduce, they crawl together and produce a slug-like blob that then sends up a stalk tipped with a spore case. The spores become gametes that later fertilize each other to become diploid amoebas again, a primitive example of sexual reproduction.

The section **Protozoa**, animal-like Protista, contains the division **Mastigophora**, the flagellated cells; **Sarcodina**, the amoeboid cells; **Ciliata**, the ciliated cells; and **Sporozoa**, internal parasites lacking a means of locomotion. The flagellates are thought to be the most primitive of the Protozoa. Most are symbionts. For example, the intestines of termites contain a species of flagellate that helps the termite digest cellulose. Sleeping sickness is caused by a flagellate. The familiar amoeba with its **pseudopods** is an example of a Sarcodina. Also in this group are the **Foraminifera** and **Radiolaria**, armored amoebas whose shells are found in limestone and chalk.

The Ciliata have cilia all over or in areas. This group is the largest of the protists. *Paramecium* swim around quickly hunting food, which they "swallow" down a "gullet" and then digest in food vacuoles. Ciliata have two types of nuclei: one is a **macronucleus**, and there are one or more **micronuclei**. *Paramecium* are capable of simple fission or sexual reproduction in which a micronucleus is traded with another, a process known as **conjugation**. *Stentor* is another very large Ciliata. It looks a little like a trombone with a ring of cilia to draw food into its mouth.

The best-known Sporozoa is the one that causes **malaria**. The life cycle of this Sporozoa (*Plasmodium*) is complex. The female *Anopheles* **mosquito** picks up the *Plasmodium* gametocytes when sucking blood. These gametocytes fertilize each other, and the diploid zygote forms a one-celled cyst in the mosquito's gut. This cyst produces sporozoites that migrate to the salivary glands where they are injected into the next person bitten by the mosquito. The sporozoites become lodged in red blood cells where they multiply many times causing damage to the red blood cell. The symptoms of malaria result from this damage. Some sporozoites become gametocytes, and the cycle begins again.

## Fungi

Fungi include **yeast** and **mushrooms**. They are multicellular heterotrophs. Some are **saprophytic**—they eat dead organisms—and some are **parasitic**—they eat live organisms without killing them. Their cells often have missing partitions that form long, multinucleated, hair-like projections called **hyphae**. The hyphae tangle together in fuzz balls called **mycelium**. The cell walls are made of **chitin**, also found in exoskeletons. Mushrooms are saprophytic. They secrete enzymes from their hyphae and then absorb the nutrients. The parasitic Fungi, like **ring worm** and **athlete's foot,** send hyphae into the host to absorb nutrients. **Mildews** are a parasitic Fungi that attack plants and crops. The **potato blight,** which caused starvation and a major migration of people to the United States, was caused by a fungus.

The division **Zygomycota** contains the familiar **bread mold,** which germinates from a spore sending its hyphae along the surface of the bread (stolons) and also into the bread (rhizoids). Bread molds may reproduce sexually or asexually. In **asexual** reproduction, the stolons send up **sporangiophores**—stalks with **sporangia** on the tips. The spores are released, and new molds start from each spore. In **sexual** reproduction, the hyphae of two plants touch. Nuclei from each haploid plant migrate into each tip to form gametes. The gametes fuse to form a diploid zygote. Meiosis occurs and a spore stalk is produced releasing haploid spores. Another division of Fungi is **Basidiomycota,** which includes the mushrooms. Most mushrooms live underground as their tangled hyphae feed on decaying material in the soil. When two hyphae meet, a typical mushroom will form.

## Animalia

**Subkingdom Parazoa.** Sponges have always been classified as animals, but they have some distinct differences. Thus, their phylum (**Porifera**) is put into a separate subkingdom—**Parazoa,** meaning "beside the animals." It has been debated whether sponges are just protozoan colonies. Most agree that multicellularity evolved from the protists at least twice. Sponges are **asymmetrical** and have no organs. Each type of cell seems to do its work independently. It is hard to think of a bunch of cells working independently as a tissue. The **collar cells** have flagella that create currents pulling water in through the openings in the many **pore cells** in the covering of epidermal cells. These pore cells are the basis of the name of the phylum. The collar cells have conical cytoplasmic projections (collars) around the base of their flagella. The collars help capture food from the water flowing past. The combined current from each pore exits the sponge through an **osculum.** Excess food captured by the collar cells is scavenged by **amoeboid cells,** which wander around in the wall of the sponge using the food themselves and distributing the rest. **Spicules** of calcium carbonate or silicic acid combined with protein fibers form an endoskeleton.

**Subkingdom Eumetazoa.** Eumetazoa, which makes up the rest of the animal kingdom, is divided into three sections: **Radiata, Protostomia,** and **Deuterostomia.** The Radiata are represented by two phyla: **Cnidaria** and **Ctenophora.** The Cnidaria include the **jellyfish.** These animals are **radially symmetrical** and have definite tissues. They do not have organs. Their outer layer is formed of **epitheliomuscular cells,** sensory cells, gland cells, interstitial cells, and cells that produce **nematocysts.** The nematocysts shoot out barbed, hair-like projections with poisonous tips. There is a **nerve net** capable of sending signals in either direction. The **ectoderm** is distinct from the **endoderm,** and there is a poorly developed third layer, the **mesoglea,** between them. Cell differentiation is not complete. There are examples of cells performing more than one function. For instance, epidermal cells also act as muscle cells. Food is killed by the poisons on the nematocysts lining the tentacles and pulled into the opening of the **gastrovascular cavity,** which has only one opening. **Sea anemones** are also members of this phylum. In many ways, they have the anatomy of a jellyfish standing on its head.

**The coelom.** Taxonomists look for basic features to help them classify living things. The structure of body cavities are a basic feature. The **coelomata** body plan has a body cavity that is completely lined by **mesoderm**. **Pseudocoelomates** have a cavity that is only partially lined. Portions of the inner ectoderm are covered by mesoderm and portions of the outer endoderm are covered, but neither is completely covered. **Acoelomates** are solid; they have no body cavity. Their digestive cavity is surrounded by endoderm. The endoderm is covered with solid mesoderm. And the ectoderm covers the mesoderm.

**Protostome versus deuterostome.** Another basic characteristic used by taxonomists has to do with how the first opening in the **blastocoel** stage of the animal develops. The hollow ball (blastocoel) invaginates on one side and becomes a two-layered animal with a hole on one side. So far, it looks like a jellyfish. The next step is for a hole to develop on the other side; then it will have two openings—a mouth and an anus. If the first opening becomes the animal's mouth, then the animal is called a **protostome**. If the first opening becomes an anus and the mouth forms second, then the animal is a **deuterostome**.

Protostomes have some other features. When the first division of their zygote occurs, the fate of the two cells is determined. One will form one part of the animal, and the other will form the rest. This is called **determinate cleavage**. In the early stages of division, the cells on one level fit in the grooves of the cells below. This is called **spiral cleavage**. One more basic characteristic of protostomes is that their coelom forms when a mass of mesodermal cells split. Earthworms, mollusks, and insects are examples of this kind of animal.

Deuterostomes are different in each case. At the division of the zygote, the fate of the two cells has not been determined. If the cells are split, then twins develop. Also, the cells during early cleavages line up in rows and columns. The deuterostomes are described as having **indeterminate, radial cleavage**. The coelom of a deuterostome begins as a pouch and grows to fill the space between the endoderm and ectoderm. Humans, dogs, and birds are deuterostomes and, surprisingly, so are starfish.

## Protostomia

**Protostomia** contains two dozen phyla. A few of these will be considered here. The symmetry of these phyla is bilateral. The characteristic used by taxonomists to divide them is based on the type of coelom that they have, among other things. The three phyla of acoelomate protostomes are considered to be the most primitive. The middle group of protostomes, which have three to eight phyla, depending on which taxonomist is describing them, all have a pseudocoelom. Almost half of the protostomes belong to the upper group having a true coelom. This last group includes the familiar clam, octopus, earthworm, lobster, and insect.

**Protostomes without coela**. The most interesting phylum in this group is **Platyhelminthes**, or flatworms. The planarian is an example. These are **free-living**, freshwater animals, one to two centimeters long, with one opening into their gastrovascular cavity. This cavity is branched many times to reach all areas of the animal's body. Two interesting features are their water balance system and their nerve system. **Flame cells**, so named because their beating cilia look like flames when seen through a microscope, are found at the ends of a network of tubules. Water and some waste enter the tubules through the flame cells and then exit the animal through pores at other locations in the tubule network. They have a **bilateral nervous system** with a system of cross connections between the two longitudinal branches. The "brain" is located in the front. This is perhaps the simplest animal with bilateral symmetry. Having two sides means that there must be a front (head) and a rear. Most of the planarian's sensory organs are in the front, which is an advantage to an animal that crawls forward. Two **eyespots** with the ability to sense the direction of light are on the head.

Two other classes of flatworms include the **flukes** and the **tapeworms**. Both of these classes are **parasitic**. Flukes have thick cuticles covering and protecting their bodies from the enzymes found in their **host**. Parasites have *degenerated*, or gone backwards, evolutionarily.

They often do not have sensory organs, muscles, digestive systems, and so on. The tapeworm is the primary example of this lack. It does not need to do anything except reproduce. It has a **scolex,** or head, with suckers and hooks to hold on to the intestine of its host. It lives by absorbing the nutrients around it. Its body is composed of repeating sacs of reproductive organs called **proglottids**. As the proglottids form behind the scolex, they mature. Their eggs are fertilized and grow. Furthest from the scolex, the ripe proglottids separate and exit the animal with its feces. If the appropriate host eats food contaminated with tapeworm eggs, the eggs hatch in the digestive tract and hook on to the intestinal wall.

Flukes, which belong to another flatworm class, have more complicated life cycles. **Schistosomiasis**—a disease that afflicts 200 to 300 million people in tropical regions near slow-moving water—is caused by the parasitic blood fluke called a schistosome. The adult worm enters a person's skin and digests its way into a blood vessel. Eventually, it finds its way to the tiny blood vessels of the intestine and lays its eggs there. The eggs grow and hatch, breaking the blood vessel. Beginning with a cough, body pains and a rash, this affliction develops into a situation where the host is so weakened that he or she may die. The eggs exit the person and develop into ciliated larvae in snails, reproducing asexually in the snail and eating the snail. They leave the snail and swim out to attack the next person wading in the water.

**Protostomes with pseudocoela**. Two phyla from this group will be mentioned: **Rotifera** and **Nematoda**. The **rotifers** are microscopic, not much bigger than a single-celled protozoan. They have a flame-cell excretory system. A wheel of beating cilia surrounds the mouth and sweeps food into the animal, which anchors itself with a **foot** while it eats. Rotifers have a complete digestive tract with a mouth and an anus, along with specialized organ systems.

The **roundworms** (Nematoda) are thin, tapered, and covered by a cuticle. They have only longitudinal muscles and thrash about inefficiently. Also called **nematodes**, these little animals are abundant. A bucket of pond water or a shovelful of garden dirt may hold a million of them. Some of them attack food crops. Another is well known as causing **trichinosis,** which forms cysts in muscle tissue and is passed by eating undercooked meat. More than fifty species of roundworm parasitize humans, including: **ascaris,** an intestinal parasite; **hookworms; pinworms;** and **filarial worms,** which can block a person's lymph system.

**Protostomes with coela.** A third of the protostome phyla have true coela, which arise from a splitting of a solid mesoderm. They also have well-developed organ systems. The phylum **Mollusca,** the second largest phylum in the animal kingdom, is in this group. Generally speaking, this phylum shares some physical features—first, a large, **ventral foot;** second, a **visceral mass,** which contains the organs; and third, a **mantle,** which covers the visceral mass and secretes a shell if the animal has a shell. The mantle often encloses gills as well. Mollusks have an **open circulatory system.** The blood is pumped to open sinuses where it washes over the tissues and then is picked up by the heart to be squirted out over the tissues again.

One of the most primitive mollusks is the **chiton,** which is bilaterally symmetrical, ovoid in shape, and has eight plates across the back for protection. These mollusks crawl around slowly on their foot, scraping up algae with a hard-toothed, tongue-like organ called a **radula.**

The class **Gastropoda** includes snails. The familiar garden snail with its coiled shell is an example. This class also includes animals with reduced shells or no shell. Their larva are bilateral, but asymmetry often develops in the adult. Most of the gastropods live in the ocean. The slug-like **nudibranchs**—some well camouflaged, others brightly colored—are common near shore. The abalone, conch, periwinkle, and limpet belong to this group.

Members of the class **Bivalva** have two shells, which can be clamped together tightly for protection, and do not have radula. They are **sessile**—stick to one spot—and they are **filter feeders.** One remarkable bivalve, the **scallop,** is able to snap its shells together and propel itself through the water. It also has numerous eyes around the edge of its mantle, and it constantly pumps water over the **gills** and over filters to remove food particles.

The class **Cephalopoda** includes octopi, squid, and chambered nautiluses, which are specialized for the rapid, coordinated motion necessary to catch prey. They have well-developed eyes that are similar to our own. Squid have developed an internal, cartilaginous skeleton and even a skull, and they have an extremely well-developed nervous system. Squid have been caught that have tentacles up to fifty feet long and that weigh two tons. Octopi are able to learn, and they are escape artists that can wiggle out of the smallest crack in a tank.

**Annelida**, the phyla of segmented worms, have a well-developed digestive system. They exchange gases through their skin or through gill-like extensions on each segment and have a closed circulatory system. A pair of protonephridia in each segment functions like human nephrons. Annelids have an enlarged front ganglion (a brain) that is linked by a parallel, double-nerve cord to a pair of ganglia on the floor of each segment, which controls their complex muscle arrangements in each segment. And they have a hydrostatic skeleton.

The class **Polychaeta** are marine annelids. They have well-developed heads with antennae and eyes. In the free-swimming species, each segment behind the head has a **parapodium** on each side that aids in swimming and functions as primitive lungs. Some polychaetes live in tubes that they secrete. In these stationary types, the body is simplified—no parapodia—and large feather-like structures develop that the worm waves out the end of the tube to catch things and to exchange gases. A lot of polychaetes have specialized sex organs located in certain segments. Each segment has many bristly **setae**, which give this group its name.

The class **Oligochaeta** includes the earthworm. It has only a few setae on each segment. The earthworm is often chosen to represent the segmented worms. **Hirudinoidea**—leeches—have specialized suckers at each end that allow them an inchworm kind of locomotion. Many of this group are bloodsuckers that have mouths in the anterior sucker. Once they have cut or dissolved their way through an animal's skin, they secrete an anticoagulant (hirudin).

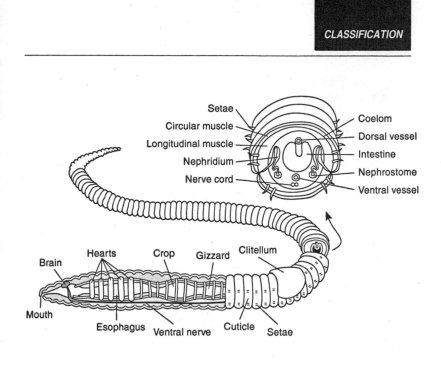

Setae
Circular muscle
Longitudinal muscle
Nephridium
Nerve cord

Coelom
Dorsal vessel
Intestine
Nephrostome
Ventral vessel

Brain
Hearts
Crop
Gizzard
Clitellum

Mouth
Esophagus
Ventral nerve
Cuticle
Setae

■ Figure 31 ■

The phylum **Onychophora** is a small but important group because it shows a combination of annelid and arthropod characteristics. These animals live in moist areas under the rotting vegetation of a forest and are a little like caterpillars with legs. They have a cuticle and nephridia like annelids do, but they also have claws and an open circulatory system like arthropods.

**Arthropoda** has the greatest number of species of any phylum. Taking into account the diversity and the ability to survive in the greatest number of habitats, Arthropoda should be ranked the most successful phylum ever. Most of the arthropods are insects. The major characteristics of this group are their chitinous exoskeleton and jointed legs. The muscles of arthropods are striated and as complex in organization as those of humans. The nervous system is like the annelid's system. The brain is dorsal and connected to a double ventral nerve cord. Instead of having ganglia in each segment, though, arthropods often have several larger ganglia.

Arthropods' sensory abilities are excellent. Their compound eyes are well developed, and they are able to sense sound waves. After all, insects make noise to attract mates. Their prospective partners must be able to hear. They also have a well-developed hormone system. Some send out **pheromones**—air-borne hormones—to communicate or to attract. The circulatory system of Arthropoda is open, with a **hemocoel**, or body cavity derived from the blastocoel, containing blood that sloshes around. In aquatic types, the gills are well developed and protected under an extension of the shell. Those living in the water get rid of their ammonia wastes through their gills.

The subphylum **Trilobita** is a well-known fossil arthropod, an ancestral arthropod lacking one thing. Trilobites' appendages are not specialized in any way but seem the same in each segment. There are no trilobites alive today. The subphylum **Chelicerata** shows a great deal of specialization in appendages. The chelicerates have bodies divided into two parts, a **cephalothorax** and an **abdomen**. There are no antennae on the head, but the anterior appendages are specialized as mouth parts called **chelicerae**. These are pincer-like structures or fangs. The abdomen has no paired appendages. The cephalothorax has up to ten appendages, which function as walking legs.

In some groups, the front legs develop into **pedipalps**, which are long and act like feelers and tasters. **Horseshoe crabs** are part of the subphylum. The Chelicerata called **Arachnida** are the most common—the spiders and ticks. These have simple eyes and two body parts. Their appendages constitute a pair of chelicerae, a pair of pedipalps, and four pairs of walking legs. In **spiders**, the chelicerae are fangs connected to a poison sac. Spiders also have silk-producing glands on the rear tip of their abdomens. An interesting adaptation in spiders is their **book lungs**, a chitin-lined chamber containing a series of leaf-like plates for gas exchange. The spider draws air in and out of this chamber by muscle action. Scorpions have a longer abdomen with a specialized segment on the end containing a poisonous spur.

Members of arthropod subphylum **Uniramia** (lately, the **Mandibu-lata**) have **antennae** and **mandibles**, two things that the Chelicerata do not have. Many of them have **maxillae**, appendages also modified as mouth parts, which manipulate the food and hold it for the mandibles.

The mandibles are modified for biting and chewing or sometimes for piercing and sucking. The class **Crustacea** includes lobsters, crabs, shrimp, barnacles, and sow bugs. The group is amazingly diverse. Characteristics true of all of them are two pairs of appendages, a pair of mandibles, and two pairs of maxillae.

Two classes of Uniramia that seem similar are the **Chilopoda** and the **Diplopoda**. Chilopoda are **centipedes** with one pair of legs per segment. The first pair of legs is poison-tipped, and they are predatory. The Diplopoda, or **millipedes,** have two pairs of legs per segment and are plant and debris eaters. They are not closely related to centipedes, and so they have a class of their own. Both have **Malpighian** tubules for excretion, and both breathe through a tracheal system. These points will be explained further in the section on insects.

The class **Insecta** occupies every niche on the earth except in the sea; it is a highly diverse and successful group. Insects have three body parts: **head, thorax,** and **abdomen.** The segments of their heads are fused. Insects have three pairs of mouth parts. An upper and lower lip—the **labrum** and **labium**—have been added to the pair of mandibles and a pair of maxillae. **Compound eyes** and one set of antennae are prominent in many. Each unit of the compound eye faces a different direction. Depending on the insect brain's ability to perceive, their view of the world may be a mosaic of pixels.

A few fossil insects showed up during the Devonian period. They were among the earliest animals on land. During the Carboniferous period, insects began to seriously diversify. Fossils of dragonflies with two-foot wing spreads have been found. Flowering plants began to diversify and spread during the Cretaceous period, and the insects enjoyed another bloom of evolution. That bloom may be nearing its end as the present human-dominated world loses its diversity of habitats.

Many insects are fliers. The wings are not appendages that have been adapted like other insect appendages or even like the wings of birds and bats. Insect wings seem to have evolved from an evagination of the upper thorax. To move their wings, insects must move their whole thorax. They have three pairs of legs from their thorax. The nerve system of insects is annelid-like; however, many of the ganglia have fused so that the thoracic ganglia can be larger than the brain and there can be fewer ganglia in the abdomen. This may be advantageous because the legs and wings are attached to the thorax near the large ganglia. Even some of the sensory organs are on the legs—touch, taste, and hearing. The "brain" may only steer the insect. Many of the insect's instinctive abilities are stored in its thoracic and abdominal ganglia.

The excretory system of insects is an interesting diverticulation of the hindgut called a Malpighian tubule. These tubules extend into the abdominal area around the hindgut and extract waste from the blood in the hemocoel. As this waste moves toward the hindgut, water is reabsorbed. In the rectum, even more water is absorbed so that the waste itself (**uric acid**) is very dry.

In insects, gas-exchange surfaces are subject to drying out and must be protected. The **tracheal system** is the respiratory system of insects. Along the sides of the insect are tiny openings called **spiracles**. These openings are connected to a branching network of tubules leading down into the animal's body. Gases diffuse easily into these tubules, and when the insect is exerting itself and needs more oxygen, the flexing of its body hclps draw air in and out.

## Deuterostomia

The protostomes are a vigorous and diverse group, but somewhere before fossil records were left, another group split off from them. This second group, the deuterstomes, had their heads at the other end of their bodies (the blastopore became an anus), their coela began as bubbles of mesoderm from their endoderm, and their initial zygote cleavages were indeterminate and radial. Protostomes and deuterostomes do share bilateral symmetry (predominantly), the same DNA mechanisms, the same metabolic chemistry, and the same mitochondria.

**Echinodermata**, a phylum that includes the **starfish, sea urchin, sand dollar, anemone,** and **sea cucumber,** is an unlikely candidate for being the phylum most closely related to ours, but it has all the same differences from the protostomes that we do. The key similarity is that their larvae are bilateral. Apparently, the **radial symmetry** of the adults was an adaptation to their originally **sessile**, or immobile, life-style. These stationary, radially symmetrical ancestors of the echino-derms were able to face the world no matter which direction it came from. The more mobile forms developed later. The name of this phylum means "spiny skin." The **spines** are projections of the **internal skeleton**. The sea urchin carries this spininess to an extreme.

The starfish is a typical echinoderm with bumps instead of spines. Its five-fold radial symmetry is like that of all the echinoderms. Its mouth is under the central disk, and its anus is on top. Each arm contains digestive glands and gonads. Its coelom is branched and serves as a circulatory system. Protected by the spines or bumps are numerous tiny projections of vascularized tissue, which serve as gills. The sexes are usually separate. There is a ring of nerve tissue around the disk that sends nerves into each arm. Any of the arms is able to be the head and each arm ends in an eyespot. Their epidermis is loaded with taste and touch sensory neurons.

A unique feature of starfish is their **water vascular system**, a series of canals in each arm connected to a ring canal in the disk. The starfish is able to exert pressure or suction on the canals. The canals end in rows of **tube feet** that line the bottom of each arm. The tube feet can extend and wave around, and when they touch something, they can attach by suction. And while holding an object, the tube feet can be shortened.

The phylum **Hemichordata** shares all of the deuterostome charac-teristics and shows the link between Echinodermata and Chordata. The **acorn worm** is an example. The key feature is that it has **pharyngeal slits**. The water goes in through these slits, over the gills, and then out through **atrial pores**. These gill slits and the pharyngeal pouch link the hemichordates and chordates. The feature that links them to the echinoderms is their larvae. Both have a unique-looking larva, found nowhere else in the animal kingdom, called a **dipleurula larva**.

The last phyla to consider in the animal kingdom is **Chordata**, which is divided into three subphyla: the **Tunicata** (or **Urochordata**), the **Cephalochordata**, and the **Vertebrata**. What unites these animals is that at some stage in their lives they all had a dorsal **notochord**; a pharyngeal pouch; a hollow, dorsal nerve cord; and a tail that extended past the anus. The larval tunicate has all these features, but when it settles down to become an adult, it loses its bilateral symmetry, dorsal nerve cord, and notochord but retains its pharyngeal pouch. A common variety is called the **sea squirt** because when you bump it, water squirts out as it tries to shrink up and hide. They attach to docks and breakwaters and are found in tide pools.

The subphylum **Cephalochordata** is best represented by a little animal called a **lancelet**. This animal has all the features of the Chordata phylum. It differs from the vertebrate subphylum mainly in that it does not have vertebrae in its back. It retains the notochord throughout its life. Its well-defined pharyngeal pouch pulls water in through the mouth by the action of cilia. Food is trapped in the pharyngeal slits, and gas exchange occurs there, also. The water is then expelled through holes along the side. The development of gill arches from the pharyngeal structure and the development of jaws from the gill arches can be followed in living animals and in fossils.

The subphylum **Vertebrata** contains all animals with backbones in addition to the other deuterostome characteristics. The most primitive vertebrate is the **Agnatha,** or jawless fish. **Lampreys** belong to this group. Instead of jaws, the lamprey, a two-foot long, eel-like fish, has a sucker lined inside with teeth. It attaches itself to its prey by suction and then bores into its prey's body to get at the juices inside. Ancient Agnatha called **ostracoderms** were armored and may have eaten plants or debris on the floor of the sea. The ostracoderms gave rise to three kinds of fish—sharks, bony fish, and lobe-finned fish—that are alive today.

Sharks are in the class **Chondricthyes** and are characterized by their **cartilaginous skeleton**, a soft skeleton seen as a specialization. The hard-boned fish are in the class **Osteichthyes** and differ from the sharks by having hard bones and a swim bladder. These animals "rule" the waters today.

This whole line of animals began their existence in fresh water and then developed ways to cope with the osmotic problems of sea water. Sharks entered the sea first. Their solution to the water-balance problem was to be isoosmotic by retaining urea in their blood. Osteichthyes made the move much later. They remained hypoosmotic to the sea. To keep from bloating up with water, they have short loops of Henle in their kidneys and a weak salt concentration gradient between the cortex and medulla of their kidneys. This adaptation allows them to excrete copious amounts of dilute urine. The lobe-finned fish, also an Osteichthyes with hard bones, was not as competitive in the open water. The lungfish and coelacanths found today are examples of this type of fish, of interest because land animals developed from this line. The Devonian—the period of the transition of life onto the land—was a time when lungfish were common. It is interesting to note here that lungs developed in fish before swim bladders did. The lobe-finned fish, left behind to flop in the mud between drying pools of water, gave rise to the terrestrial animals.

Members of class **Amphibia** lay fish-like eggs with no amnion and fertilize these eggs externally like most fish. Their larval stage is fish-like, and finally, after a metamorphosis, their adult stage is somewhat adapted to land. The amphibians have legs, sensory systems, and respiratory systems that function well in air, but they are unable to stand hot, dry conditions. During the Carboniferous period, they bloomed into an environment without much competition. However, the upheavals that marked the end of the Permian period wiped out all the amphibians, except the ones surviving today—salamanders, toads, and frogs.

Members of class **Reptilia**, which displaced the amphibians, are completely adapted to living out of water. Some notable examples—crocodiles and turtles—have returned to the water but still have the terrestrial adaptations. The most major of these adaptations is the development of the **amniotic egg**. This egg is waterproof, contains fluids in which the reptile embryo can grow, and is protected by a leathery shell. The "land egg" and the internal fertilization that accompanies it makes it possible for the reptile to live its whole life on land.

The reptiles also have skin that is waterproof, unlike amphibian skin. The legs of reptiles are stronger and better suited for holding the animal off the ground. Amphibian hearts are three-chambered; they have two atria and one ventricle. Although well suited to breathing through the skin, this type of heart would involve too much water loss on the land. In a reptile's heart, the ventricle is mostly divided by a septum, which separates blood from the lungs and blood from the body. Reptiles have a bone structure better able to resist gravity—for example, a bigger, stronger rib cage. Reptiles have been very successful. Despite some memorable extinctions, they number over 6,500 species today. The living examples are **turtles, crocodiles** and **alligators, snakes** and **lizards,** and the rare **tuatara.**

It is important to remember that mammals, dinosaurs, crocodiles, snakes, and turtles all originated from ancestral amphibians. During the time of dinosaurs, mammal-like reptiles called **therapsids** were present. The therapsids had hair, differentiated teeth, and legs under their bodies but were unable to compete with the dinosaurs. Mammals did progress to small, scurrying, insectivorous creatures during the Jurassic period. In fact, some have speculated what would have happened if dinosaurs had not become extinct. It is likely that mammals would still be in the background. The reason for this extinction is debatable. There is good evidence that a meteorite or a group of meteorites struck the earth and caused a "nuclear" winter. Whatever happened, a change in the environment definitely occurred that the dinosaurs were not able to cope with. In our world, birds and crocodiles are the closest relatives to dinosaurs.

Two kinds of dinosaur became expert flyers. The **pterosaurs** had thin, webbed wings and were successful for a long time. Fossil pterosaurs range in size from a few inches up to huge examples with sixty-foot wingspreads. They must have been successful in a number of habitats. The other dinosaur experiment with flight involved scales turning into feathers. Some well-preserved samples of **Archaeopteryx** show an animal that is anatomically a dinosaur except for the imprint of feathers in the fine-grained rock. Some taxonomists insist that birds are dinosaurs.

The class **Aves** includes all birds. They are warm blooded. Their feathers are a unique feature found nowhere else. Feathers are light, are good insulators, and make the bird aerodynamic. Birds have four-chambered hearts with a complete division between ventricles and differ from mammals in that their aorta arches to the right. Their lungs are also different. Whereas mammal lungs dead end in alveoli, the lungs of birds are open at both ends. The air goes through the bird's lungs into air sacs in its body and then back through the lungs. The air sacs are an unusual feature that also helps to streamline the body. Another adaptation for flight are bones that are hollow and webbed inside. Birds also have highly developed senses, especially visual, and they have interesting behaviors that center around mating, nest maintenance, and rearing of their young.

The class **Mammalia** is noted for its ability to produce milk. Mammals are warm blooded, are covered with hair for insulation, and have larger brains than do reptiles. In fact, the behavior of reptiles and birds seems more controlled by inherited responses (**instincts**) than by **learned behavior,** as it is in mammals. Mammals don't lay eggs, except in the case of the **monotremes,** who still have a few reptilian characteristics but do have hair and do provide milk for their young.

The other two types of mammals are the **marsupials** and the **placentals.** Examples of the marsupials are the kangaroo and other pouched animals. Their babies are born in an immature state and must climb from the vagina to the pouch on their own, find a nipple, and stay there while they complete their development. Placental mammals have been the most successful. Their young are kept in the uterus until they are well developed. A placental mammal baby requires a great deal of care but grows quickly. There are four major groups of mammals. Bats and shrews are the most like the ancient insectivores. A second group includes rabbits, horses, pigs, sea cows, elephants, and porpoises. The third group is composed of the carnivores. And the fourth group includes monkeys, apes, and humans, all belonging to the **primates**.

## Algal Protists

Some important distinctions between the plant kingdom and the algal protists are that the large majority of algae live in water, have no distinct cell differentiation, have no roots, stems, or leaves, and have no support or conductive tissues. Their reproductive cells have no sterile jacket surrounding and protecting them. And finally, the zygotes of algal protists leave the female structures of the plant before developing into embryos. The three divisions of the algal protists are based on the color of their pigments.

**Chlorophyta.** The **Chlorophyta** (green algae) are those from which plants have evolved. They have chlorophylls *a* and *b* and carotenoids. The chloroplasts of green algae have grana-like structures. The most primitive is *Chlamydomonas*, a one-celled alga without cellulose that lives as a scum in fresh water. It has two flagella and a single cup-shaped chloroplast. Like plants, *Chlamydomonas* has a life cycle that involves **alternation of generation**. *Chlamydomonas* is haploid except for a zygote step in its life cycle. While it alternates between sexual and asexual reproduction, most of the time, it reproduces asexually. A cell absorbs its flagella, and two, four, or more cells form by fission within it. The new cells regrow their flagella and break out. They are called **zoospores** until they grow to the size of the adult cell. Sexual reproduction starts with the adult vegetative cell absorbing its flagella and mitotically producing four to eight gametes, which develop flagella and swim off. These gametes have the ability to recognize another appropriate gamete, and the two fuse to form a diploid zygote, which divides meiotically to produce four, haploid daughter cells, which grow to become adults. Because the gametes look alike, this process is called **isogamy**.

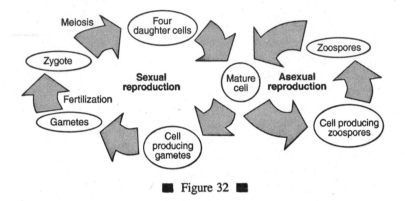

■ Figure 32 ■

The evolutionary trend toward multicellularity can be traced in the green algae. Some species are colonial. *Gonium* is one of the most primitive. It forms a plate containing several cells, which look like *Chlamydomonas* and are held together by a mucus covering. Cytoplasmic strands connect the cells and coordinate swimming. *Pandorina* is a more complex colony, a hollow ball with some primitive differentiation. It swims in one direction, and if separated from the colony, individual cells die. Its sexual reproduction is **heterogamous**; the gametes are two different sizes.

This trend toward increasing size and specialization reaches its height in *Volvox*. The colonies can have up to a half million cells, all tied together by strands of protoplasm in giant, hollow balls. Most of the cells are vegetative, but some cells in the lower portion of the sphere are reproductive. One type of cell develops many motile gametes (**sperm**), and a female cell produces a relatively large **egg**. This is **oogamous** reproduction. The sperm swim over to the egg and fertilize it. The zygote goes through meiosis to become a new colony. The trends suggest (but do not prove) an evolutionary sequence from single cell to multicelled organisms. Organisms get larger. The cells in these organisms have some communication. There is a definite division of labor as cells specialize to become vegetative or reproductive. And there is a progression from simple isogamy to heterogamy.

*Ulva*, another green alga, shows some advanced traits. It has a large, lettuce-leaf-like structure, two cells thick; it lacks conductive tissue; its life cycle shows a good alternation of generation with a multicellular body in both diploid and haploid stages. Beginning with fertilization, the zygote develops mitotically into a multicellular diploid plant. This plant grows **sporangia** in which meiosis produces flagellated zoospores that each grow into a haploid plant. The haploid plant produces gametes in **gametangia** that fuse to form diploid zygotes, and the cycle is complete.

In the alternation of generation so far, the haploid phase has been dominant, and only the zygote is diploid. In *Ulva*, a diploid phase is apparent. The haploid multicellular plant is called a **gametophyte**, and the diploid multicellular plant is called a **sporophyte**. The gametophyte produces gametes, and the sporophyte produces spores.

**Phaeophyta and Rhodophyta**. The **Phaeophyta** (brown algae) are the most impressive and largest members of the kingdom Protista. The giant seaweeds of this group can grow to 300 feet long in offshore forests. The leaf-like structures are called **blades**. The stem is a **stipe**. And the point of attachment to the sea floor is called a **holdfast**. These algae have no need for conductive tissue because they are bathed in an isoosmotic solution. They differ from the Chlorophyta by having **fucoxanthin**, a brownish pigment; having chlorophyll *c* instead of chlorophyll *b*; and storing energy in a chemical called **laminarin**. The seaweeds that we are familiar with feel slippery because the cells secrete **algin**, a gelatinous material that cushions the cells and keeps them from drying out. The brown algae have true alternation of generation with both gametophyte and sporophyte represented.

The red color of **Rhodophyta** is due to an accessory pigment called **phycoerythrin**. Red algae differ from green algae in other ways. They store their energy in a glycogen-like **floridean starch**. Their cell walls secrete a material like algin, and they often have chlorophyll *d*. Remember that all multicelled (nonbacterial) autotrophs have chlorophyll *a*, but only Chlorophyta and members of the kingdom Plantae have chlorophyll *b*.

## From Water to Land: Plants

There are three problems in moving from a water to a land environment:  gravity, drying out, and gamete transportation.  Gravity creates a need to be strong and stiff enough to stand up.  In upright plants, the problem of transporting water and nutrients up and down the stem exists.  Drying out can be solved by a waterproof covering, but then how can gas exchange occur without moist membranes being exposed to the air?

Plants differ from the algal protists by having a sterile jacket of cells around their multicellular reproductive organs.  The **antheridia** contain the male reproductive cells in a sterile jacket of cells, and the **archegonia** contain the protected female reproductive cells.  The egg cells stay in their protective covering, and the sperm travel to them.  This process is oogamous.  The embryo begins its development in the archegonium safe from a hostile environment.  Even the spores are produced in multicellular structures protected by a covering of sterile cells.

**Bryophytes.**  The plant kingdom is divided into two divisions, **bryophytes** and **tracheophytes**.  The first terrestrial plants—the bryophytes —were low growing and lived close to water in shady places.  Included in this group are mosses, hornworts, and liverworts.  They lack vascular tissue, and most important, they still produce flagellated sperm cells and must be wet to reproduce.  The evolutionary trend toward a larger sporophyte is not seen in bryophytes.  Their sporophyte grows from the gametophyte, has no chlorophyll, and is dependent on the gametophyte.  Bryophytes do not represent a transition between algal protists and higher plants but are a side branch that has made its own adaptations from an ancestor common to both.

The plant called moss by most people is haploid.  On the tops of some moss stalks are archegonia, and on the tops of other moss stalks are antheridia.  Sperm produced in the antheridia is splashed over to a nearby archegonium, into which it swims and fertilizes the egg.  The diploid zygote divides mitotically into a stalk that grows out of the archegonium. This diploid stalk grows a sporangium at its tip to produce haploid spores. Each spore grows into a **protonema** that becomes a moss gametophyte.

Liverworts—for example, the *Marchantia*—are prostrate, leaf-like structures with **rhizoids**, root-like structures, projecting from their lower surfaces. Again, the green, dominant part of their alternation of generation is haploid. The diploid sporophyte grows entirely within the archegonium. The antheridia and archegonia are located in tiny structures resembling miniature palm trees. Another interesting trait in *Marchantia* (and other liverworts) is the formation of little cups called **gemmae cups**. Asexually produced clusters of cells called **gemmae** can be splashed some distance from these cups by raindrops.

**Tracheophytes.** The plants in this division have developed vascular tissues that move water and nutrients throughout the plant and woody tissue that is stiff enough to hold them up against the pull of gravity. There are five subdivisions in this group, here arranged from the simplest to the most complex. The first four subdivisions are **seedless**; only the fifth subdivision, Spermopsida, produces **seeds**. The classification presented here is a logical one, but other schemes exist.

1. **Psilopsida:** The most common is called *Psilotum* or **whisker fern**, whose fossils come from the Silurian period, 395 million years ago. Whisker ferns are leafless plants with dichotomously branching, vertical, green stems. They grow from underground stems (rhizomes) which have single-celled, hair-like rhizoids projected into the soil.

2. **Lycopsida:** This subdivision first appeared in the late Devonian over 340 million years ago. The group has a vascular system and therefore true roots, stems, and leaves. During the Carboniferous period, giant lycopsids grew over a hundred feet tall and had trunks six feet across. Their leaves were **emergent**, which means that they stuck out from the stems. They died out when the swamps dried up. The survivors are small

and **herbaceous** (nonwoody). The sporangia of these plants grow on specialized leaves called **sporophylls**. Often the sporophylls are arranged in cone-like clusters known as **strobili**. The club shape of the strobili leads to the name **club mosses**, although they are not a moss. An interesting development occurs in *Selaginella*, the appearance of specialized strobili that produce **megaspores**. These larger spores develop into female gametophytes that produce only archegonia. Also, there are strobili that produce microspores that grow into male gametophytes. A plant like *Selaginella* that is able to produce male and female spores is said to be **heterosporous**. Plants that produce spores that can be either female or male gametes are **homosporous**.

3.  **Sphenopsida**: This subdivision became prominent in the Carboniferous period, but only one genus survives today. The survivor is called **horsetails**. Much of today's coal is from these plants. Sphenopsids have a vascular system. Horsetails are homosporous. The gametophytes of this group are very small; the leaves arc emergent in whorls around the stem at intervals.

4.  **Pteropsida**: These are the **ferns**. They were important plants during the Carboniferous period and survived the changes that killed off most of the other subdivisions. It is thought that the fern developed from the psilosids. The fern leaf may have formed when the branches flattened and began to fill in between with a web of vascular tissue, which has the advantage of presenting a greater surface area to the sun for photosynthesis. This type of leaf is found on all seed plants today and is called a **megaphyll**. The ferns have excellent vascular systems. In the tropics, ferns grow twenty feet tall. Ferns are homosporous. Their haploid gametophytes are flat, heart-shaped, and lie close to the ground with rhizoids on their bottom surface. They have chlorophyll and live independently but can survive only in moist habitats. The antheridia send out sperm which swim toward the archegonia. When there are many gametophytes overlapping each other, the sperm of one

may swim to the archegonium of another. In the archegonium, the egg is fertilized and grows out and upward. At first, this sporophyte is dependent on the gametophyte, but then it sends out rhizoids and begins an independent life. The fern **frond** (leaf) unfurls as it grows and has been called a **fiddlehead**. When the frond matures, rows of specialized structures called **sori** develop under the frond. The sori protect clusters of sporangia. These sporangia are able to fling the spores away from the fern under the proper conditions.

5.  **Spermopsida (the seed plant):** The most important development in plants was the **seed**. The seed is an embryo sporophyte (diploid) in suspended animation wrapped in a protective coat with a food supply. Seeds may grow in places impossible for a spore to germinate. In the Spermopsida, the sporangium and its structures not only protect the megaspore but hold and protect the female gametophyte that grows in it. The microspores, which become **pollen,** are produced elsewhere on the sporophyte. They blow or are carried to the specialized structures of the sporangium containing the female gametophyte. A pollen grain becomes a male gametophyte that grows toward the female gametophyte. Neither gametophyte is exposed to the environment. Both are protected in the structures of the sporangium. Fertilization occurs in the sporangium, and the zygote develops along with a specialized supportive and protective structure until it is a seed. Only then does the sporophyte open up and release the seed.

There are six classes in the subdivision Spermopsida. The first five classes are informally called the **gymnosperms,** and the sixth class is called the **angiosperms.** All of these produce seeds. The earliest seed plants are found in fossils dated from the end of the Devonian period, 360 million years ago. The forests of the swampy Carboniferous period were filled with examples of Lycopsida, Sphenopsida, Ptcropsida, and early gymnosperms. The amphibians and the first reptiles must have adapted to take advantage of the seed as a new food source.

## The Life Cycle of the Conifers

These plants are informally called the **conifers** because they bear cones. They are represented by firs, cedars, yews, cypresses, hemlocks, larches, redwoods, spruces, and pines. The trees are evergreen and are most common in the northern hemisphere at high altitudes or upper temperate latitudes.

The life cycle of the conifers is highly advanced. The **pine** life cycle starts with a seed. The seed germinates into a large diploid plant—the tree. Two kinds of cones develop on the mature pine tree: the **ovulate cone**, which contains sporangia to produce the megaspores, and the **pollen cone**, whose sporangia produce the microspores. Meiosis occurs in each of these cones to produce the heterospores. The female cone is composed of a densely stacked pile of **scales**. On each scale is a pair of sporangia covered by a tough layer of cells. A **micropyle** is a small hole through the covering aimed toward the center of the cone. Meiosis produces four megaspores, and three of these are absorbed. The fourth haploid megaspore undergoes many cell divisions to produce the **megagametophyte**. This female gametophyte gives rise to two or more archegonia near the micropyle, each with an egg. The whole structure consisting of the covering, the sporangium, and the female gametophyte is called the **ovule**.

Pollen cones are dense spirals of sporophylls, each bearing lots of sporangia. These produce millions of microspores, which become the male gametophytes. The male gametophyte divides to become four cells wrapped in a tough coat and sporting two short wings. Only two of these four cells survive to become the tube cell and the generative cell. This four-celled structure is the pollen grain. Because pine trees are wind pollinated, millions of pollen grains are released, but only a few ever reach the next stage of the life cycle.

Those lucky few fall into the spaces between the scales of a female cone and get stuck in a sticky liquid secreted from the micropyle. The female cone, by the way, has been developing for a year and opens itself at the right season to catch the wind-borne pollen. After **pollination**, it closes back up. By opening this way, it exposes the ovule; thus, it is named **gymnosperm** (naked seed).

The liquid dries up, pulling the pollen in through the micropyle. Contact with the sporangium causes the male gametophyte (the pollen grain) to resume growing. It sends out a **pollen tube,** which contains the nucleus of the **tube cell.** Close behind is the **generative cell,** which divides into a **spermatogenous cell** and a **sterile cell.** The sperm cell divides again to become two sperm cells. The tube cell grows through the wall of the sporangium to the archegonium. One of the sperm cell nuclei can now fertilize the egg. The diploid zygote divides mitotically to become an embryo pine tree. The female gametophyte is a food supply for the embryo. The ovule is now a seed, and the pine's life cycle is complete.

## Angiosperms: The Flowering Plants

Flowering plants appeared 140 million years ago, after the start of the Cretaceous period. While the dinosaurs ruled the animal world, angiosperms became dominant in the plant world. Sixty-five million years ago, the dinosaurs had all gone, but the flowering plants continued to spread and to diversify. Along with this spread of flowers during the Cretaceous period came a whole new group of insects and mammals.

**Flowers** are the distinguishing reproductive structure of angiosperms. All flowers have the ability to produce pollen and/or seeds. Some plants produce *incomplete* flowers, which are either pollen producers or seed producers. The others are *complete* and have pollen and seed all on the same flower. A typical complete flower sits on a **receptacle,** the end of a stem that holds a flower. The **sepals** cover the **bud.** They open to allow the **petals** to emerge. All the petals together are called the **corolla.** The corolla and sepals form the **perianth. Stamens** emerge just inside the corolla ring. These hold the **microsporangia** in **anthers** at the end of a long stalk called a **filament.** At the center is the **carpel.** It has a sticky **stigma** raised up on a long **style** that is connected to an **ovary.** The ovary may be divided into compartments called **ovules,** which are the megasporangia. There may be one or several carpels fused together to make a **pistil.** The parts of the flower have evolved from four rings of leaves (sporophylls) whorled around the end of a stem. The rings became the sepals, petals, stamen, and pistil.

The modifications to this basic layout are many. For flowers that are wind pollinated, the corolla is unnecessary. Instead, the stamens are multiplied so that clouds of pollen may be produced. Flowers pollinated by bees are sweet, open during the day, and are usually in the yellow to blue end of the spectrum because bees can see into the ultraviolet. Flowers pollinated by moths are white and open at night, with a "heavier" smell.

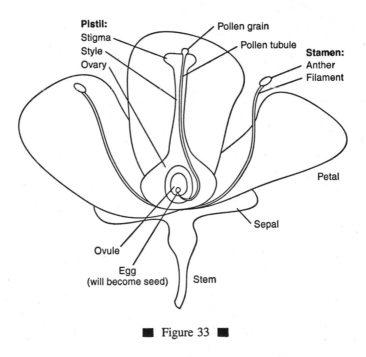

■ Figure 33 ■

## The Life Cycle of an Angiosperm

The seed grows into the flowering plant that produces flowers. In the anther of the stamen, microsporangia produce haploid microspores that become pollen grains. In the ovule, the megaspore divides to form a seven-celled structure that has a large, double-nucleated (**polar nuclei)** cell in the middle with three small cells at one end and three small cells at the other end. The middle, small cell at one end is the **egg cell**. The whole ovule containing this seven-celled female gametophyte is attached to the wall of the ovary by a stalk and has a small hole (a micropyle) near its attachment to the stalk.

When the pollen grain reaches the stigma, it becomes the male gametophyte consisting of a long pollen tube. In the tube, the tube nucleus leads the way, followed by two sperm nuclei. When the tube reaches the ovary, it grows to the micropyle and then through the wall of the ovule to fertilize the egg. The second sperm nucleus fertilizes the large, middle cell making it a triploid cell. This triploid cell grows into a food supply called **endosperm** for the embryo. Meanwhile, the fertilized egg is becoming an embryo. The seed has formed. At no time is the seed ever exposed to the air; it is always locked up in the ovary. If this were a pine tree, the scales of the cone would now separate, and the seeds on their wings would blow away. But this is an angiosperm, and the seeds are still enclosed in the ovary. The ovary, which contains many seeds, is called a **fruit**, another unique distinction of angiosperms. The fruit can be fleshy and nutritious if animals are the means of its dispersal, or it can be hard and covered with hooks if it is to be caught in animal fur for dispersal, or it can be light with some kind of frothy umbrella if it is to be dispersed by the wind.

**Evolutionary trends of plant development.** The major trend has been the increasing dominance of the diploid sporophyte in the life cycle of the plant. From the early chlorophytes through the bryophytes, pteropsids, conifers, and angiosperms, the gametophyte became more protected. During this development, homospory was replaced with heterospory.

In becoming successful on dry land, plants lost their swimming sperm and developed ways to protect themselves from drying out. They had to have a good vascular system. The seed was a big improvement because it was diploid, had the advantage of two sets of genes, was already somewhat developed, and had its own food supply. The seed is better than a spore.

## Anatomy of a Plant

**Monocot versus dicot.** The class **Angiospermae** is divided into two subclasses—**Dicotyledoneae** and **Monocotyledoneae**, which are informally referred to as **dicots** and **monocots**. There are several distinctions between them. The most obvious is that the **veins** of dicot leaves are usually **webbed,** while the monocot leaves have **parallel** veins. Also, dicots have floral parts that come in multiples of four or five, while those in monocots are in multiples of three. Dicot roots are usually in the form of a **taproot,** while monocots have **fibrous roots.** The stems of dicots have their fibrovascular bundles in **rings** around the trunk. Monocots have their bundles scattered around. And finally, the embryo plant in the seed has either one or two leaves called **cotyledons.** A dicot has two embryo leaves, and a monocot has one embryo leaf. The cotyledons are used as a food supply while the roots and foliage develop.

**The seed.** Seeds can remain **dormant** for years until the conditions are right for them to sprout, which is an important adaptation. Some desert seeds must have a minimum amount of rain to end their dormancy, enough to insure that the fast-maturing plant will have enough water to grow and produce its own seeds. Some unusual seeds in chaparral regions have to be heated by a brush fire before they germinate. These plants grow well in the ashy soil. Some seeds must pass through an animal's digestive tract. Others must be scratched and tumbled in a stream.

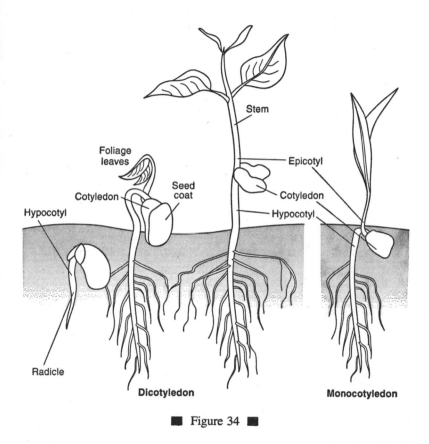

Figure 34

For most seeds, **germination** involves **imbibition**, or soaking in enough water to make the seed swell and break its coat. This rehydration stimulates the production of enzymes that "wake up" the embryo. The first thing to poke out of the seed is the **radicle,** which is the root of the embryo. The radicle is continuous with the **hypocotyl**—the projection "below the cotyledon." In many dicots, the radicle grows downward, and the bent hypocotyl lifts the seed above the ground. Here the seed coat is discarded, the embryo leaves spread, and the

**epicotyl** extends in a stem bearing the first leaves. In monocots, the process is similar. The radicle is the first thing out. The hypocotyl sends a shoot called a **coleoptile** to the surface. The coleoptile has a leaf wrapped up inside and unfurls the leaf above the ground.

**The roots.** The roots anchor the plant, help hold it upright, and absorb water and minerals from the ground. Roots may also act to store food. The busiest part of a root system is the **root tip**. The **apical meristem** just under the **root cap** is a region of cells undergoing mitosis. The root cap is pushed forward through the soil particles. Behind the apical meristem is a zone of **cell elongation** where new cells grow. Next is the zone of **cell maturation** where the cells achieve functional maturity. Here, **root hairs** extend from the root to absorb water.

The root tip is a structure made of three concentric tubes. From the outside ring inward, these are the **protoderm, ground meristem**, and **procambium**. The single layer of cells in the epidermis comes from the protoderm. Root hairs extend from this epidermis, greatly enhancing its surface area. Water and minerals pass into or through the epidermis. Next is the ground meristem, which gives rise to the tissue filling the **cortex**. The procambium forms the **stele**, a central tube containing the vascular tissue of the plant—the **xylem** and **phloem**. There is a one-cell-thick layer called the **endodermis** between the cortex and the stele. The endodermis cells are glued together by a **Casparian strip**. Just inside the stele is a layer of cells called the **pericycle**, which becomes **meristematic**—able to divide—and can give rise to secondary roots.

Water in the soil is in contact with the root hairs. Because these are hyperosmotic, water is absorbed, making the root hair cytoplasm slightly more dilute—hypoosmotic—than the next cortex cell inward. The passing of water continues into the stele as the cells near the xylem are the most hyperosmotic. The tough cell wall allows great osmotic pressures to build up without risk of rupturing the cell. Water and its dissolved minerals and salts may also be channeled into the stele by soaking along the hydrophyllic cell walls of the cortex. End-to-end xylem cells form a vessel through which the water is initially pushed up the stem.

**The stem.** Stems hold plants up, conduct photosynthesis, act as places to store food, and transport water and sap. A **vascular cambium** separates the cortex from the pith in a dicot. This cambium divides to renew itself and to form phloem cells outside and xylem cells inside. In a woody dicot, this growth leads to the familiar **rings** that mark annual growth. In a **herbaceous**, nonwoody dicot, the structure is not as clear. And in monocots, there is no easily seen distinction between the cortex and the xylem. What is obvious are the **fibrovascular bundles.** They have large xylem cells surrounded by thinner phloem cells, all wrapped in a ring of protective cells.

**The leaves.** A leaf consists of a **blade** attached by a **stipule** to a stem. When a leaf prepares to fall off, a layer of cells near its attachment to the stem and called the **abscission layer** weakens and breaks. Leaves come in all shapes, and there is a name for each shape. A **chordate** leaf is heart shaped. A **lanceolate** leaf is shaped like a spearhead. An **oval** leaf is shaped like an oval. Leaves may be **simple** if they have one margin that is not indented clear to the midrib or **compound** if their indentations reach the midrib. The leaves could even be **double compound** if their leaflets are divided into more leaflets. The margin of a leaf may be **smooth** or **serrate**. The leaf is **lobed** if the margin is deeply indented. The venation of the leaf may be **pinnate**, or feather-like, with one main vein up the middle and secondary veins branching off from that. Or the leaf may have **palmate** venation with a few equal veins branching out from its stipule. The characteristic venation of a monocot is **parallel.**

The cross section of a leaf reveals that its upper surface is covered by a **waxy cuticle** on a layer of **upper epidermis** cells. The cuticle prevents water loss. Between the upper and lower epidermis is the **mesophyll** layer. Here are the cells that are most actively engaged in photosynthesis. In some leaves, the mesophyll cells closest to the upper epidermis are arranged like columns or palisades. These are called **palisade mesophyll**. Below them is the **spongy mesophyll**, a collection of cells jumbled on top of each other with air spaces between them. Running through the mesophyll are the **veins**. The veins are wrapped in **bundle sheath cells** and contain phloem and xylem. On the bottom of the leaf is the **lower epidermis**. Like the upper, the lower epidermis is composed of flat cells that run edge to edge like a cobblestone street. Here and there is an opening called a **stoma**. Many of these openings are called **stomata**. Another type of leaf may have basically the same structures, but instead of a pallisade mesophyll, it has these cells wrapped like a wreath around the veins.

The stoma must open to allow gases in and out of the leaf. Photosynthesis depends on taking in carbon dioxide and getting rid of oxygen. The membranes of the mesophyll must be moist to allow gas diffusion. The problem is that if the stomata are open all the time the leaf will dry out. The mechanism controlling the opening and closing of the stomata is easily explained. The stoma is the opening between two cells called **guard cells,** which contain chlorophyll. These cells have a thick, stiff, cell wall on the side of the hole and a flexible and stretchable cell wall away from the hole. When the guard cells become hyperosmotic to the surrounding cells, they draw water into themselves and swell up. Because their inner surfaces cannot stretch, they arch away from each other like the reflection of an arched bridge over a pond. When they are hypoosmotic, they shrink, and the stoma closes.

One theory to explain this behavior is that in their hypoosmotic state guard cells have stores of insoluble starch. When the dawn light hits the guard cell, photosynthesis begins, and the dissolved carbon dioxide is used up, raising the pH and making the cell less acidic. An enzyme that breaks down starch is present in the guard cell. This enzyme is inactive in an acidic medium but active in a neutral medium. When the light strikes the guard, this enzyme becomes active as photosynthesis makes the medium neutral and changes the insoluble starch to glucose. The addition of glucose to the medium makes the guard cell hyperosmotic, and it swells up causing an opening to appear between it and its partner guard cell. Other theories are being researched, but researchers agree that flaccid guard cells close the stomata, and turgid guard cells open the stomata.

**Getting water up the trunk**. The tallest tree in the northwestern United States is a gymnosperm that is 364 feet tall. How does it conduct water all the way to the top? First, there is root pressure. **Root pressure** is created by the movement of water into the stele. The osmotic pressure in the roots can lift water several feet. On a humid morning, water drops can be seen exuding from the ends of the veins in herbaceous leaves. This process is called **guttation** and is due to root pressure. But the water isn't up far enough. The ability of a tree to pull water the rest of the way up is called the **transpiration-tension theory**. The inner surface of xylem cells is hydrophilic. Water is attracted to these surfaces. Also, water molecules form hydrogen bonds with other water molecules that can be thought of as long chains. Now, imagine the uppermost water molecule at the top of a plant diffusing through the xylem into the mesophyll of the leaf. This molecule is bonding to the next one downward. When it goes into the leaf, the second water molecule takes its spot. Because the water molecules are connected in a chain, the whole chain moves up one link. Losing water from a leaf is called **transpiration**. This transpiration creates a negative pressure that puts **tension** on the chains of molecules leading back to the base of the plant and pulls water up the xylem tubes.

## Introduction

Ecology is the study of the interactions between living organisms and their environment. Anything other than the individual is part of the individual's environment. Five levels of ecology are **population, community, ecosystem, biome**, and **biosphere**. **Population** is a group of individuals of the same species. **Community** is a group of populations interacting in the same area. An **ecosystem** is a community plus all the physical factors that affect it, like the amount of sunlight, the humidity, the type of soil, and so on. A **biome** is a world-sized ecosystem—for example, desert or rain forest. The **biosphere** is the whole, thin film of life on and near the earth's surface.

## Population

**Distribution.** There are three ways that a population may be distributed over an area. The first is a **random** distribution, which would occur only if all parts of the environment are equal and the dispersal of seeds or young animals is not affected by any factor such as competition or food supply. This situation is rare. However, many sampling techniques used to estimate the population numbers in an area do assume a random distribution. The accuracy of these sampling techniques depends on the assessor's ability to discern patterns of distribution before making an estimate.

Another distribution pattern is **uniform**. In this situation, the environment is uniform, and there is strong competition. This competition prevents young individuals from being able to exist near established individuals. This pattern of distribution shows up occasionally. An example is the distribution of creosote bushes in deserts. Each established bush secretes chemicals into the soil that prevent another plant's roots from growing too close. In another example, during breeding season, the males of many species will mark a territory and defend it from other males. If the males in an area are all evenly matched and the food sources are evenly dispersed, then the animals will be roughly evenly spaced. Another seasonal example is the nests of shore birds that are built just out of pecking range of each other.

The third type of distribution is **clumped**, with the clumps, or aggregations, spread out randomly. This is the most common distribution of plant and animal populations; it assumes that not all areas of a range are equal. Some places may have better **microclimates**—areas with more water, protection from the wind, or better soil. In a territorial species, the stronger will claim the better places and attract more mates. Competition will be more intense in favorable spots among both plants and animals, but the rewards offered by the better site will compensate. A clump of trees is better off in a strong wind than is a single tree. A herd of zebras has an advantage over a solitary zebra.

**Population sampling techniques.** The **quadrate** method maps out a checkerboard that covers an area and then randomly selects a few of the squares. An estimate of the population can be made by counting the individuals in the selected squares and calculating the proportion that would be in all the squares. If 10 squares out of 100 had 32 animals, then the total population is estimated to be 320. Another variation is to make a one-meter square frame and throw it randomly onto a field. The number of beetles found in the square meter multiplied by the square meters in the field gives an estimate of the beetle population. The person doing the assay must make judgments about where to throw. Results can vary from person to person. The bigger the area counted, the more the frame is tossed, the more accurate the results will be.

Using a **line transect** works well on plants. You throw or shoot a rope in a random direction and then measure the length of each plant of a species under the rope. This data can be used to calculate the percentage of coverage of the area for that plant. Another method is to draw a straight line on a map and then hike along the line. Each plant within a meter of your path is counted. This technique leads to some interesting, and occasionally difficult, hikes.

Estimates of population size may be made indirectly by counting burrows, footprints, or droppings. Another method is called **mark and release**. A number of traps are set, and the captured animals are marked and released. Assume that 100 were marked. The same number of traps are set later. If 100 animals are caught the next time and 10 of these were marked, then the first 100 must have been 10% of the total. In this case, the total estimated population would be 1,000.

The population data is referred to as the **density** and can be expressed as the number of organisms per area. If the habitat is three-dimensional, the population may be expressed as the number of individuals per volume. This knowledge helps scientists keep track of increases and decreases and allows better planning in terms of land management.

An animal or plant may be an **indicator species**. An indicator species may not be obvious to the casual observer, but it may have links to the whole ecosystem which will allow ecologists to judge the health of the whole ecosystem. An indicator species is like a thermometer in the mouth of a patient. So when a snail darter fish or a spotted owl in the Oregon forests makes the headlines, it is not the animal but rather its whole ecosystem that is the issue.

**Growth and regulation.** The **biotic potential** is a measure of the reproductive potential of a species. A pair of oysters can produce millions of offspring at a time. Even a pair of elephants producing one or two offspring each year for twenty years can rapidly build a large population. What prevents the world from being overrun by elephants or anything else is **environmental resistance**. This resistance takes many forms, and most are **density dependent limiting factors** (DDLFs). As the density of a population increases, the DDLFs take an increasing toll. Disease is a DDLF. If a population is crowded together, diseases spread through it faster than if the animals or plants were more spread out. Predators are DDLFs because they will eat rabbits when rabbits are everywhere. The food supply is a DDLF because it runs out faster when there are more animals to feed. The crowding of mice in a lab disturbs their endocrine balance so much that their breeding rate drops.

At some level, a population will exceed the **carrying capacity** of the area. When there are too many individuals and not enough space or not enough food, the DDLFs cut back the numbers, and the death rate exceeds the birth rate. Soon the population density will dip below the carrying capacity, and the DDLFs are not as effective at reducing the numbers. The population may start to increase again until it is over the carrying capacity. In a stable population, this fluctuation of the population density above and below the carrying capacity is a normal thing. If one group starts to increase uncontrollably, then the whole ecosystem is thrown out of balance.

The human population used to be stable. Neanderthal and then Cro-Magnon peoples may have reached 30 million. By the time that the bubonic plague was decimating Europe, the human population in the world may have been than 300 million people. That's a ten-fold increase in perhaps 20,000 years. There has been another ten-fold increase in the last 500 years. The carrying capacity of the earth is undefined. Technology keeps providing and providing, but technology uses up the fields and the forests of the earth in order to work its miracles.

## Community

The community is defined as the interactions within a cluster of populations. A community is usually defined by the most visible plants within it. Cacti are an **indicator species** for a desert.

Each species of plant in a community lives independently of other plants and within its range of physical needs, including temperature, rainfall, and soil type. When there is a sharp change in physical conditions in a given direction, the boundaries of a community in that direction will be more clearly defined. This may occur at the boundary between original soil and erosion-deposited soil or between the south-facing side and the north-facing side of a hill.

**Properties**. One of the properties of a community is its **diversity** of species. There are also other factors, like the **equitability** of numbers. This is a measure of the relative size of each population in the community. Another measure is the **importance** of a species to the community. Wolves may not be as numerous as the larger game that they prey on, but they have a major influence. A bacteria in the soil may not be noticed, but plants in that community may depend on it.

**First impressions** are important. The abundance and height of vegetation and its structure and type are all clues to the nature of the community. **Energy distribution** is a feature ecologists look for. In other words, who eats whom. The **stability** of a community can be measured by how consistent the population densities remain over a period of time.

**Interactions**. Interference with a community may have unexpected results. In Borneo, the World Health Organization (WHO) set out to eliminate disease-carrying mosquitoes by spraying with DDT. They did, and the thatched roofs of the villages began to rot and cave in, which was certainly unexpected. It was found that moth larvae were responsible. Their predator, the animal that primarily kept their numbers low, was a parasitic wasp that laid its eggs in the larvae. The wasp was sensitive to DDT; the moth was not. A sudden rat infestation was another undesirable effect. Along with the rats came the diseases that they helped to spread. Also, geckos continued to eat insects exposed to DDT, including the moths. The geckos began to die as their internal DDT levels increased. The gecko, by the way, is a big-headed lizard that makes a barking sound from which its name is derived. Cats, which eat geckos as well as rats, were also killed by the DDT in their diet. Thus, the environmental resistance affecting rats was reduced, and its biotic potential caused its numbers to increase to a new carrying capacity. To fix the problem, WHO turned to other malaria controls besides DDT and had the Royal Air Force parachute hundreds of cats into the villages in a procedure called "Operation Cat Drop."

The key species on which this whole community hinged was the parasitic wasp that kept the moths in check. A community is a carefully balanced system that can be unbalanced with interesting results. It used to be thought that the more complex the interactions of a community, then the more stable that community would be in the face of pressures like this. It has been found that simpler communities do react faster, but they also recover their original levels faster. A complex community reacts more slowly to an upset, but the echoes continue to reverberate for a long time. A rain forest, one of the world's most complex communities, is just as precariously balanced on its key environmental factors as is a simpler community like a grassland.

**Relationships. Predator-prey** is an obvious relationship. Lions are the predators, and zebra are the prey. Less obvious is the relationship between a mosquito and a reindeer on the arctic tundra. Or the relationship between a person and a cow when that person is in a supermarket picking out dinner. **Predation** includes these and other cases. Predation involves sensing, finding, stalking, catching, and eating.

**Symbiosis** is the "living together" of two different species; there are three types of symbiosis. **Parasitism** is the case where one of the species benefits, and the other is harmed. A successful parasite cannot be hurt by its **host** nor does it usually kill its host. **Parasites** are usually examples of **degenerate evolution**. They have lost many organ systems that they would need if the host did not provide everything. They specialize in producing vast numbers of eggs and having complicated life cycles. **Mutualism** is the condition where both animals benefit. Termites could not digest the cellulose in wood if it were not for an intestinal protozoan that does this for them. Both the termite and the protozoan benefit. **Commensalism** is the case where one individual in the relationship benefits, and the other does not care. A nest in a tree is an example. The tree is not affected by the nest, but the bird benefits greatly.

## Ecosystem

The ecosystem is the community including all the physical factors that affect it. When ecologists talk of an ecosystem, they are talking about everything. The physical factors of an ecosystem are called **abiotic factors**. They include the slope of the land, the type of soil, the amount of rainfall, the average temperatures, the length of seasons, the nearness to bodies of water, and the velocities of wind, to name a few. Many of these may be called **density independent limiting factors (DILFs)**. A field of gophers wiped out by an early snowfall are all affected whether their population is dense or sparse.

The **habitat** of an animal or plant describes closely where the animal lives. Giraffes inhabit an open, hot plain with tall trees and occasional water holes. The habitat of a polar bear is the open ice of the Arctic Circle. A complete description of an animal's habitat would include the temperature range that it can tolerate and what kind of terrain and plants and other animals that it lives with. An even more complete description of the animal would be its **niche**. The niche includes the habitat but adds what the animal does, describes what it eats, how, where, and when it hunts and reproduces, and so on.

Once the niche of an animal has been described, it becomes possible to measure levels of **competition**. Suppose a species of fox has the same niche as a cat except that it lives in old badger caves and the cat lives in hollow logs. If the number of mice is adequate, these two would not be fierce competitors. If there is a shortage of mice, the cat and the fox would compete indirectly. However, if the fox tries to live in one of the limited number of hollow logs, then there might be a direct and ferocious confrontation.

The **competitive exclusion principle** states that the closer the niches of two animals or two plants are, then the more they will compete with each other. The result of the competition may be the extinction of one or a sideways evolution of either or both into slightly different niches. This result is called **character displacement**. If one is superior in size and strength or is able to hide better or has a resistance to a critical disease or is just luckier, it may win.

**Energy flow.** The sun is our ultimate source of energy. It emits **wavelengths** that are the energy for life on the earth. The **magneto-sphere** and the atmosphere filter out all but radio, heat, light, and ultraviolet waves. Most of the UV light is stopped by ozone high in the atmosphere. Green plant cells do not just absorb light; they convert light energy to chemical energy by **photosynthesis.**

The **greenhouse effect** occurs because carbon dioxide is transparent to light but opaque to heat waves. As the carbon dioxide percentage in the atmosphere rises little by little, the air becomes more and more of a blanket. The amount of energy that we get from the sun equals the amount of heat that the earth radiates back into space. If the amounts of energy received and radiated were not equal, the earth would cool off or warm up. The greenhouse effect and **global warming** are being enhanced by our burning of fossils fuels with the resultant carbon dioxide increase in the atmosphere.

A **food chain** describes the flow of energy in an ecosystem. The sunlight is used by a plant to make energy-rich molecules. An animal eats the plant and converts that energy into its own molecules. People grow grains to feed to cows so that they can eat the cows. Anything siphoning energy from this food chain is quickly eliminated.

A **pyramid of biomass** is usually drawn as a triangle. By far the largest group are the **producers** who are placed at the bottom. These are all the photosynthetic organisms. A narrow band above them across the triangle represents the **primary consumers.** Above that in smaller and smaller bands are the **secondary consumer, tertiary consumer,** and so forth. Each of these levels is called a **trophic level.** The word **herbivore** describes animals that eat plants; they are the producers. A secondary consumer or higher is a **predator.** There are various other descriptive terms: **carnivore** (eats meat), **omnivore** (eats plants and meat), **insectivore** (eats insects), **saprovore** (eats dead things that have been discarded). An important aspect of this pyramid that gets left out is a place for the **decomposers.** These are bacteria and fungi that cause once living material to break down into its atomic components so that it can be used again.

A **food web** is created by visualizing who eats whom in a commu-nity. A beetle eats some rice in a village in Borneo. The beetle is eaten by a spider, a predaceous beetle, a bird, a toad, or a gecko. An arrow is drawn from the beetle to each of the secondary consumers. Arrows are now drawn from the spider to the bird, toad, and gecko. The cat eats any of these, and so arrows are drawn from each to the cat. A hawk or eagle could eat a cat, and so an arrow should be drawn from the cat to the hawk. What is being described is a complex interweaving of arrows that signify the passage of materials and energy from one individual to another in a food web.

**Biological magnification** occurs when materials like DDT do not dissipate from an animal's body but stay dissolved in the tissue. If a gecko eats 100 insects sick on DDT, a lot of this DDT dissolves in the gecko's tissue. Eventually, the gecko gets sick, making it easy prey for a cat. By the time a cat eats 10 sick geckos, it has the equivalent of enough DDT to kill 1,000 insects. DDT does not break down. It is not **biodegradable,** and it has the property of collecting at the top of the pyramid.

**Usable energy** comes into the ecosystem in the form of light and is converted to chemical energy. This conversion is not 100% efficient; so some of the original energy is lost as **unusable heat.** Heat cannot be used, and so it just radiates away. Energy is lost as heat at each trophic level. Decomposers work until finally all the energy that came in has been converted to heat. The amount of energy that enters an ecosystem is equal to the amount of heat that leaves the system.

**Matter Cycles.** The earth has not lost or gained a significant amount of matter since its beginning. Whereas energy flows through a system and is gone, matter is recycled. The process is similar to an overshot **water wheel.** The water comes over the top of the water wheel at a high energy level. It is dumped into the compartments of the wheel and turns the wheel as it falls. When its compartment reaches the bottom of the turn, the water is dumped out and discarded. The wheel recycles back to the beginning. In this analogy, water represents energy, in at high levels and out at low levels, and the wheel is matter, recycled over and over.

1.  **Water Cycle:** Heat from the sun allows **evaporation** from oceans, lakes, and streams. **Transpiration** lets even more water into the air. The energy of the sun also causes **convection** in the atmosphere. Combined with the rotation of the earth, convection causes **winds**. Warm air spirals upward as colder air slips in underneath. The rising air cools because of expansion. If its **humidity** is high, then some of the water vapor will condense. Condensation releases more heat to the air mass, causing it to rise more. This process of the air rising and expanding while not cooling off is called **adiabatic expansion**. The moist air eventually forms clouds, and rainfall returns the water to the ground, where it collects in lakes or in animals and plants before it is returned to the air or to the sea.

2.  **Oxygen Cycle:** Oxygen is produced during photosynthesis by the **splitting of water**. The oxygen is then taken up by a mitochondrion and used in aerobic cell respiration as the low-energy **electron acceptor**, which—along with two hydrogen ions—becomes water again to repeat the cycle. There are great reservoirs of oxygen in the earth. Vast deposits of calcium carbonate and mountains of metal oxides were formed when the earth was younger. When volcanic action releases the oxygen from these reservoirs, it joins in the cycle. When an oyster adds another ring to its shell, oxygen is removed from the cycle.

3.  **Carbon Cycle:** The carbon cycle is more complex. Because the earth is not receiving any new supplies of carbon, it must recycle what it has. Beginning with **plants**, carbon may go into the **atmosphere** as carbon dioxide from **respiration** or be incorporated from the atmosphere by **photosynthesis**. Plants send their carbon on to **herbivores** and then **carnivores**. Plants and animals all eventually surrender their carbon to the **decomposers** who return carbon dioxide to the air. Animals also respire and send carbon dioxide to the air. Plant and animal remains may become **coal, oil**, and **gas** which can be released to the atmosphere by **weathering** and **combustion**.

4. **Nitrogen Cycle:** The nitrogen cycle is a lot more complicated. It emphasizes the absolute dependence of life on the earth to **soil bacteria** and fungi. The ecosystem that occurs around the roots is called the **rhizosphere**. Soil bacteria are involved in a variety of processes. First, nitrogen compounds are given to the soil as waste or tissue. Bacteria and fungi begin the breakdown of this **organic nitrogen** to **amino acids**. Some other bacteria convert the amino acids to **ammonia**. Other bacteria oxidize the ammonia to **nitrites**. Still another set of bacteria oxidizes the nitrites to **nitrates**. Keep in mind that when something is oxidized, something else is reduced. And reduction is a process that builds bigger molecules, in this case, in the bacteria. The nitrate can be used by plants to grow and the cycle continues. There are other bacteria that conduct the process of **denitrification** to convert nitrate to **nitrogen gas**. Nitrogen gas is 78% of the atmosphere. It can enter the nitrogen cycle through **nitrogen fixing** bacteria, which prepare the nitrogen for plant consumption. **Lightning** also fixes nitrogen. Humans fix nitrogen in **fertilizer factories** and throw it on plants by the ton.

## Biome

Ecosystems can be as large and general or as small and specific as the ecologist wishes. On the large end, they reflect general consistencies that are worldwide. At this scale, they are called **biomes**. All the life at or near the earth's surface is called the **biosphere**. A few biomes follow.

**Tundra.** This area is on the top of the world and is characterized by **permafrost**. The ground may thaw only at the surface, creating drainage problems and making for very wet, mushy footing. During the summer, the daylight lasts twenty or more hours, and the fast-growing, low plants make the most of a short summer. Many animals migrate into the area. A great number of birds get fat on the abundant insect life. Caribou take advantage of the grazing opportunities. Wolves follow the caribou and the rabbits. Foxes and lemmings also are abundant. There are no tall trees here, only low shrubs and gently rolling terrain.

**Taiga.** The next biome south of the tundra is the **taiga**. This biome, characterized by evergreen trees, stretches across the top of North America and Siberia. The soil in the taiga does thaw during the summer, and there are more plants and animals in this biome. The animals associated with the taiga are wolves, moose, and bears.

**Deciduous forest.** In this kind of forest, the leaves change to brilliant hues of red, orange, yellow, and purple and fall off in the autumn. The temperature in this area is much less harsh than that in the tundra and taiga. It is warmer with gentler winters. Examples of this biome are found in the eastern United States, Europe, and in the far-eastern portion of Asia. The plant and animal diversity has increased from the tundra to the taiga and to the deciduous forest.

**Tropical rain forest.** Although scarcer by the year, **tropical rain forests** are found in the north of South America, the mid-west of Africa, across India, and down the island chain into the northeastern parts of Australia. These are the most complex and diverse ecosystems known, and diversity increases as one approaches the equator. The tundra has vast herds of only a few kinds of animals; the tropical rain forest has a few members each of a vast number of species. The tops of most of the trees are interlaced and in the full sun. This **canopy** is full of animals and plants that never set foot on the ground. The floor of the rain forest is dark and still.

**Grasslands.** The **steppes** of Russia, the **pampas** of South America, the **veldt** of Africa, and the **Great Plains** of North America are all examples of **grasslands.** The western end of the Great Plains is characterized by short grass, barely knee high. At the eastern end, stories are still told of grass taller than a Conestoga wagon. A rainfall gradient exists across the United States. In the **rainshadow** of the Rockies, there is less rain than further east. Traveling eastward, there is generally more and more rain until there is enough to support the deciduous forest. Less rain means shorter grass. This is where the antelope and the buffalo roamed.

**Deserts.** **Deserts** are the driest land of all. There is plenty of sunshine but not enough water for most plants. The plants found here are specially adapted for water conservation: the cactus, Joshua tree, and ocotillo are examples. The animals are smaller and often nocturnal. Plants have incredibly short life spans. After a chance rain, their seeds can sprout, the flowers develop, and new seeds form in several days.

## Succession

A glacier may scrape the life off the ground along its path. A fire may turn an area into ash. The stage is now set for succession. **Succession** is the process of replacing the life that was there. If the conditions are the same as before, the same life will eventually return. However, it does not come back all at once. It comes back in stages—each stage preparing the way for the next.

The classic example of succession is found at the southern end of Lake Michigan. The generalizations that apply in this case are true of most forms of succession. First, the land was scraped clean by a glacier. The glacier retreated leaving only sand at the water's edge. Debris piled up along the shore, and wind-blown dust collected. Scavengers ate what washed up, and a few blades of grass sprouted. The few plants and animals that first arrived formed the **pioneer community.**

This community extended into the water because plants and fish lived in the shallow water close to shore. As dirt collected on the beach and was anchored by the pioneer plants and as plants began to live in the water near the shoreline, they softened the environment and prepared the way for less adventurous species. The seeds of these species had been blowing about the pioneer community, but they had been unable to get started until the pioneer community had softened the conditions and more animals found the habitat to their liking and stayed.

In the area south of Lake Michigan, this second group of plants and animals is called the **foredune community** and has more diversity than the first. The **biomass** is greater. The food webs are more complex. The same thing happened in the water as sediment collected in the shallows, filling them in.

Walking southward from the shore of Lake Michigan today, you begin at a beach. Within yards, you come to the foredune community, which is characterized by grasses. Beyond that is the **cottonwood community**, which has cottonwood trees, bushes, birds, snakes, a variety of insects, and so on. The cottonwood community created conditions favorable for pine. Pines tend to grow closer together. In competition with the cottonwood, the pines did better and the **pine woods** were established. You can imagine that cottonwoods once grew where the pines are today. The pine forests are taller, quieter, more protected from the wind, and support a greater abundance of other plants and animals around them than did preceding communities.

The idea is that succession moved two ways: (1) the water became land as the plants in the water held sediment and choked the shoreline, causing it to become a **marsh**, then a **bog**, then a **field**; (2) the field became the foredune community and then the cottonwood community, which grew in areas that used to be foredune. The cottonwoods were replaced by pine woods. The pine woods were replaced by **oak woods**, and the oak woods were replaced by the **beech-maple forest**.

All of this may have taken thousands of years. Looking closely at the beech-maple forest you can find no evidence that another community is coming. The beech-maple forest is deep and complex and is here to stay; it is the **climax community** for this area; it is the community that best takes advantage of the conditions of water, sun, and temperature that are found there.

The process of succession follows several observable trends:

1. Animals and plants come and go quickly at first but in the later stages are more apt to stay.
2. The number of different species grows. Each succeeding community has more niches for a greater number of species.
3. The plant life makes better and better use of the available sunlight, and a greater percentage of light is converted to organic matter.
4. More and more organic and inorganic nutrients are held in the soil and in the plants.
5. The increase in the biomass of the area continues until it reaches its maximum in the climax stage. This is true of the organic content of the soil also.
6. The plants in succeeding communities are taller, and the vertical dimension increases. This allows for different layers to be established.
7. The larger and heavier the community becomes, the more it can protect itself from climatic conditions. Winds that may be strong at the edges are softened toward the middle. Sunlight is filtered, and the floor of the forest is in shadow. These are conditions that the community itself creates.
8. The food webs are more complex, leading to specialization and to greater efficiency in utilizing the energy available.

No real boundaries exist between succeeding communities. Each plant or animal lives within the conditions for which it is best suited. They are where they are because the conditions are right for them. However, it is possible to notice cumulative patterns and to note associations that occur with regularity. Hiking southward from Lake Michigan provides an opportunity to see the overall patterns. The concept of a climax community is the idea that changes will stop when no new plants or animals can find a niche for themselves and when the patterns become stable for those conditions.

## Human Impact

The impact of people on ecology is overwhelming. Ecologists often feel that they must hurry to record the ecology of an area before it is wiped out by human intervention. The climax community that we are rushing toward might include only people, their factories, and their food sources. People do several things that impact the ecology:

First, people simplify the food webs to produce food efficiently for themselves. Thousands of square miles of corn fields and wheat fields exist across the middle of the U.S. Anything else is discouraged. Other fields grow food for cows, which become food for people. The quickest way from dirt to mouth is the most cost effective.

Second, people dig up concentrated ores to use and then randomly spread them instead of recycling. Aluminum occurs in several places in the world. Vast mining operations occur in those places. The aluminum is purified and made into useful products that become discarded all over the world. Petroleum products are concentrated in certain parts of the world. Once this material is gone, we hope that other sources of energy can be used. Responsible people the world over are already planning for alternate, cleaner energy sources.

Third, people have not shown a great regard for the habitats around them. Understandably, putting food on the table comes before saving the rain forests. Only people removed from the immediacy of the situation can take a broader view. The study of ecology has given us information about the importance of the other life forms on the earth, and it has given us alternatives. A town may decide that filling a pond with paper mill wastes creates only a small problem in exchange for higher incomes. The study of ecology can give a broader perspective in which the magnitude of this type of pollution is seen for what it is. Ecology can show what is required to trap the pollutants and convert them to environmentally safe materials. The trick is to educate people to accept the added costs of this cleanup of industrial pollution.

It is not a question of letting people go until a natural climax community is reached, where people and what's left of the environment are in a state of equilibrium. Nothing on the earth has been able to compete with humans. Recognition of this fact is dawning all over the world as people see the effects of waste and pollution. The attitudes affecting the National Wilderness areas in the United States are a microcosm of a worldwide awareness. We used to exploit. Then we renewed. We planted more, prevented fires, and air dropped food to starving deer. Next we found multiuses for our wildernesses. Ranchers could send in cattle and sheep. Timber companies could harvest trees. Because this did not work, we preserved the wildernesses; they were shielded from all human impact. This approach is not practical either. So now the attitude is to try to define a balance in the unspoiled areas of the country. To do this, we must understand the ecology.